Contents

Mathematical Origami

Geometrical shapes by paper folding

David Mitchell

Tarquin Publications

Acknowledgments

I should like to thank my many friends around the world for the help they have given me over the years, wittingly and unwittingly, in developing the models which are included in this book. Wayne Brown in particular must be singled out. He has found something to praise in the worst of my folds and something that can be improved in the best.

My very special thanks also go to Tomoko Fuse, Bob Neale, Nick Robinson, David Brill and Paul Jackson for giving permission for their ingenious designs to be included. They remain the copyright of their inventors, but my book is so much stronger for being able to include them.

David Mitchell, Kendal.

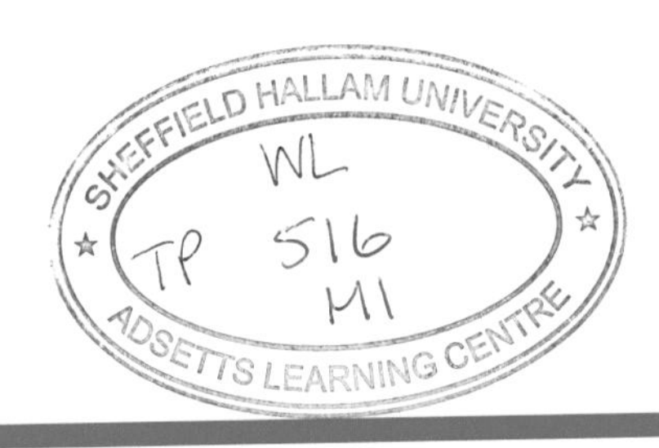

I.S.B.N.: 1 899618 18 X
Design: Paul Chilvers
Printing: Ashford Colour Press Ltd, Gosport, Hampshire

Tarquin Publications
99 Hatfield Road
St Albans, Herts
AL1 4JL
United Kingdom

www.tarquinbooks.com

Introductory Notes

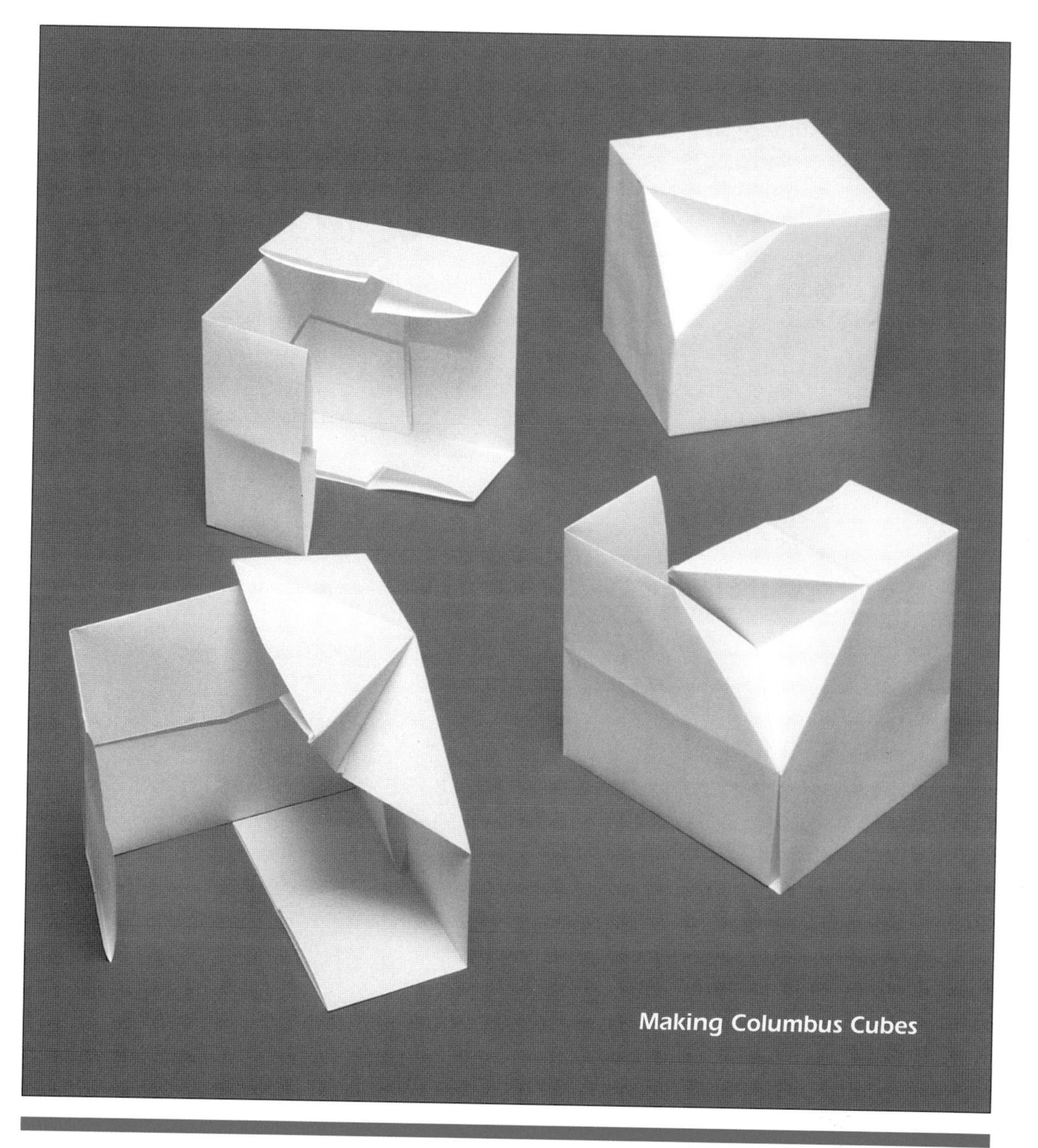

Making Columbus Cubes

Modular origami

The models in this book are examples of the relatively new mathematical art of modular paperfolding, in which a number of simple folded modules are assembled to create a polyhedral model. This kind of paperfolding originated in the United States in the melting-pot times of the early 1960's. Since then it has gained adherents in the United Kingdom and across the world, even becoming popular in Japan, the traditional home of single-sheet paperfolding, where it is known as unit origami.

As well as many new folds, this collection contains some of the finest classic models, all of which have been chosen for the simplicity and elegance of their design and the sheer mathematical beauty of the finished polyhedra themselves.

Even if you have never folded paper before you should be able to achieve creditable results first time on all the models in this collection, though you are strongly recommended to make your first two models the Cube and the Tetrahedron. In general the easy models are early on in the book and the more difficult ones towards the end.

Paper quality

Normal photocopy paper of 80gsm in weight is ideal for the models in this collection.

Paper size

The instructions for all the designs begin with standard A4 paper, which is commonly - and cheaply - available throughout most of the world. But if you don't have access to this paper size, then you can easily make paper of the same shape out of any other rectangle. The method is given as part of the instructions for the Outline Dodecahedron on page 58
One of the special properties of the A4 shape rectangle - which has sides in the proportion of 1 to the square root of 2 - is that you can cut it in half and end up with two rectangles of exactly the same shape, which you can cut in half and end up with ... ad infinitum. This makes it very easy to scale models up or down in size. The folding instructions specify the optimum starting size for easy folding, but you can make the modules however big or small the quality of your paper and your paper folding allows.

The limit of upward scaling is usually the strength of the paper. Over large models will collapse under their own weight. So if you want to build really big, try using thin card. Working small presents a different problem. As the model gets smaller the thickness of the paper soon causes the creases to distort. So if you prefer small, use the thinnest good quality paper you can find.

Cutting the paper to size

Although the instructions always begin with A4, sometimes the basis of the module is either a square or an intermediate rectangle and some of the paper has to be cut away. It is best to make these cuts with a craft knife drawn carefully along the edge of a metal rule. Most good art and craft shops now sell plastic cutting mats - which have remarkable self-healing properties - at a very reasonable price, and using one saves damaging any less suitable surface. An A4 size mat is quite large enough.

For safety always cut along the edge of the metal rule that is farthest away from you. This will usually mean turning the paper around before you start cutting, and re-orientating it to the diagrams afterwards.

How to follow the folding instructions

Paper folding instructions are set out in the form of a series of 'before' and 'after' diagrams. Each step is numbered so that the sequence of folds is clear. The symbols used in this book are based on the system invented by Akira Yoshizawa of Japan, which are now used and understood by paper folders throughout the world. There is a dictionary of the symbols and their meanings at the end of this introduction.

A typical 'before' diagram for a simple fold might look like this:

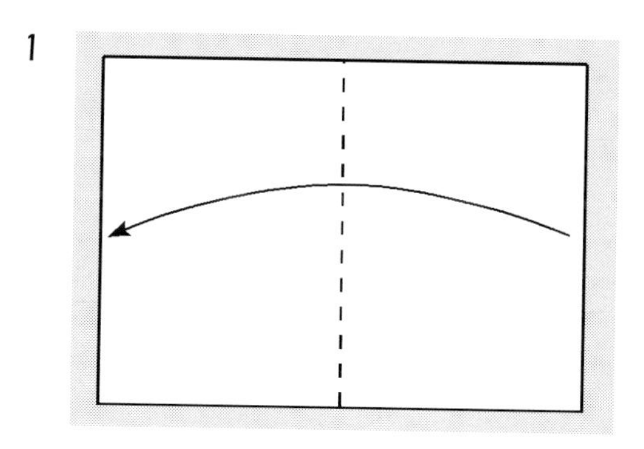

Lay the right hand edge on the left hand edge and flatten down.

In this case the meaning of the symbols is made clear by the written instruction underneath. To make this fold you would lift up the right hand edge of the paper, bring it across in the direction indicated by the arrow, and lay it down exactly on top of the left hand edge. It's a good idea to try this out. Hold the paper firmly in place and gently flatten down the centre of the fold until it becomes a soft crease. If you're happy the edges haven't moved flatten this crease out in either direction until the paper is flat, then run your fingernail up and down the new edge to really set the crease into the paper.

The result would look like this:

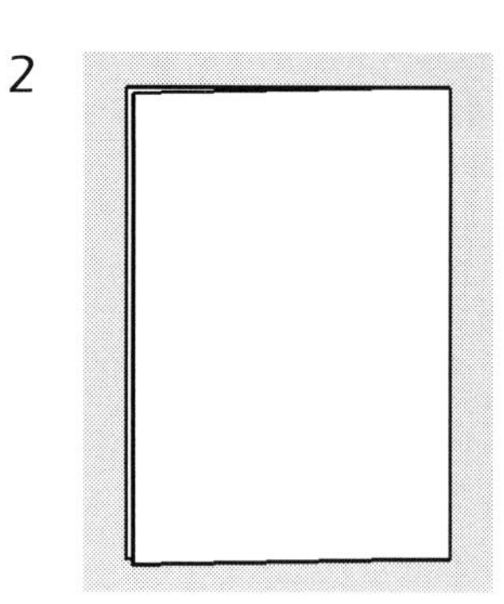

Well - almost like this. In this 'after' diagram the top half of the paper is shown slightly raised so that you can see there are two layers at the left hand edge, but if you have made the fold accurately you should only be able to see one edge on your model. 'After' diagrams are always drawn like this because you need to know there should be two edges there. If you weren't supposed to fold the two edges exactly together there would have been a special instruction about this on the 'before' diagram.

Most 'after' diagrams are also the 'before' diagram for the next fold as well. So step 2 above might really look like this:

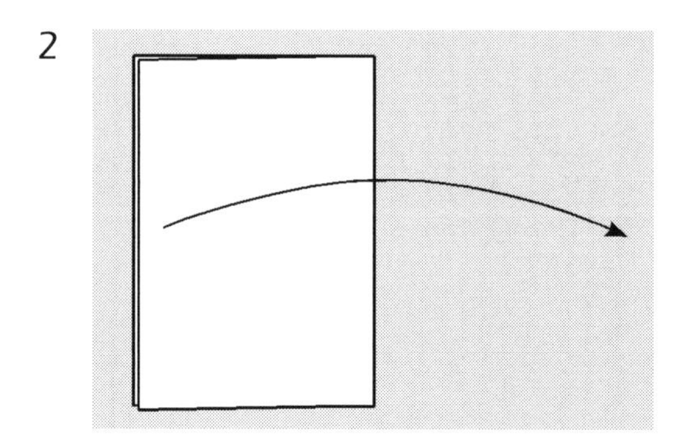

In this case there aren't any written instructions because the meaning of the diagram is obvious. It means 'Open out the fold.' Usually written instructions are saved for when the diagram doesn't give quite enough information by itself.

After you have completed this fold there will be a crease right across the middle of the paper from top to bottom. You can see how this is shown on the 'after' diagram by looking at step 3 below. Creases are always shown just after they've been made, but sometimes there are so many creases that showing them all can be confusing. Don't worry about the one's that disappear between diagrams. It simply means you won't be needing them again.

Most folding instructions simply tell you to lay one edge onto another, or a corner onto a crease, or a crease onto another crease etc. and usually its quite obvious what's required if you look at the 'after' diagram before you make the fold. But there's another kind of fold used in this book which has to be made in quite a different way. Here's a typical 'before' diagram for one of those:

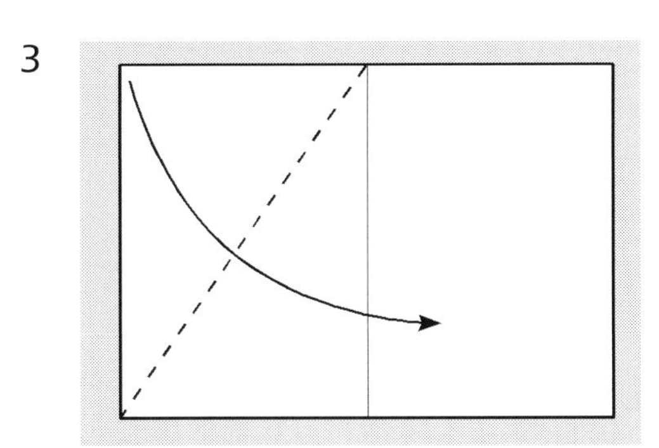

The dashed line marks where the crease will be when you flatten the paper down. In this case the new crease will run from the top of the existing central crease to the point of the left hand bottom corner. The problem is that the movement arrow ends up in empty space so that you can't know exactly where to lay the moving corner down.

The best way to make this fold is to crease it one small section at a time. Work in approximately one centimetre long sections starting from the top of the existing central crease and heading roughly in the right direction. Crease this first section down firmly but keep the rest of the fold vague until you are sure it will pass exactly through the point of the corner. Flatten it down in stages, adjusting it each time until you have it right.

This is the result:

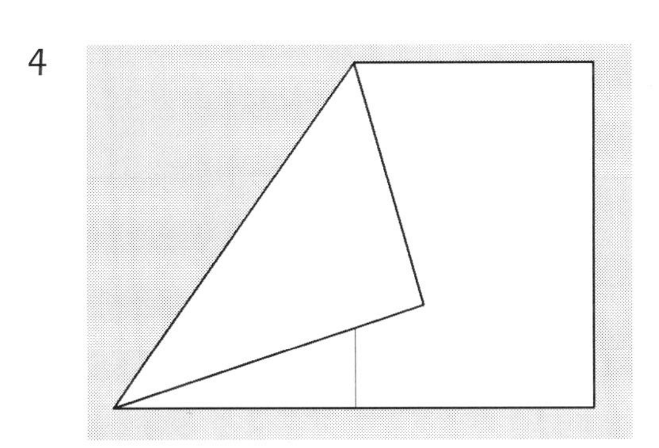

The point of the left hand corner should be sharp to the touch.

The instruction 'Turn over' is one of the commonest written instructions in the book. If you find that your model has stopped resembling the 'after' diagrams then the chances are it's because you've missed one of these somewhere along the way.

Tips on folding

▪ Its easiest to fold with your paper laid onto a hard, smooth surface. If you want to relax in the armchair while you fold then use a coffee-table sized hardback book. Folding in the air is strictly for experts or the birds.

▪ Don't be afraid to turn the paper around so that you can make the fold in the most natural way. The diagrams have been drawn to explain the folds not to tell you how to make them. What would be perfectly comfortable for a right-handed person might be torture to a left handed one.

▪ The best tools for sharpening up creases are strong fingernails, but if you have none, or if like mine they wear out through too much folding, then the curved parts of knife or scissor handles can be used instead.

▪ Each time you finish a fold work your way around the outside of the model making sure that none of the edges or corners have been pulled out of true. If they have, correct them.

▪ Treat the paper gently. Handling it roughly will cause crumpling which will spoil the clean look of the finished model.

▪ If despite this your paper goes floppy through too much handling you can restore it by opening it fully out and ironing it with a slightly warm dry-iron. The creases will be easy to refold since the lines of weakness will still be there. This is a good way of smartening up old modules but beware - too much heat will cause the edges of the paper to curl.

Tips on assembling the models

▪ Fold all the modules before you start putting them together.

▪ Be as gentle with the modules during assembly as you were with the paper while you were folding it. There is an element of puzzle involved in constructing any modular assembly. Enjoy it.

▪ The last module is always the most difficult to ease into place. If it's a case of getting a flap into a pocket then curling the flap slightly often helps. Otherwise try loosening the whole model right up.

▪ If all the modules are in place but you can't get them to fit tightly together then try just easing each module a tiny bit inwards at a time. Alternatively it may be that all the modules that meet at one point need to be pushed together in one go. Every model is different. There just aren't any rules.

Using glue

All of the basic models will hold firmly together without glue, but there are degrees of firmness. Some models are completely solid and could safely be tossed across the room. Others require rather more delicate handling. There is nothing wrong in gluing the less stable models together, but don't overdo it. If you make a fold in the wrong place you can always unfold it and try again, but once you have glued a module in place it's difficult to remove it again without destroying the whole model. For this reason it's best to use a lipstick type of glue that remains workable for a short while after you put the surfaces together, and to practice the assembly first so that you appreciate exactly how the modules go together.

Dictionary of Symbols

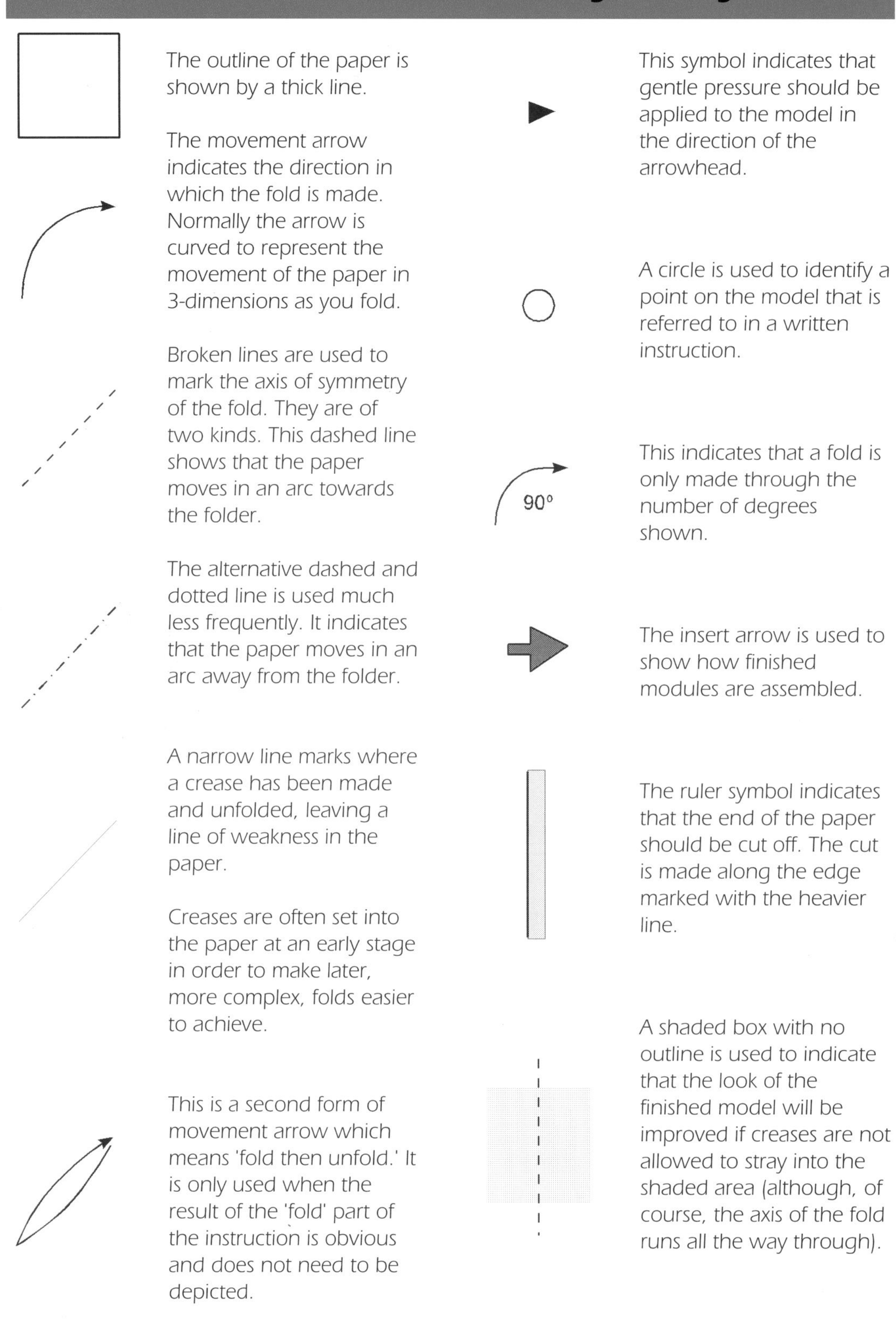

The outline of the paper is shown by a thick line.

The movement arrow indicates the direction in which the fold is made. Normally the arrow is curved to represent the movement of the paper in 3-dimensions as you fold.

Broken lines are used to mark the axis of symmetry of the fold. They are of two kinds. This dashed line shows that the paper moves in an arc towards the folder.

The alternative dashed and dotted line is used much less frequently. It indicates that the paper moves in an arc away from the folder.

A narrow line marks where a crease has been made and unfolded, leaving a line of weakness in the paper.

Creases are often set into the paper at an early stage in order to make later, more complex, folds easier to achieve.

This is a second form of movement arrow which means 'fold then unfold.' It is only used when the result of the 'fold' part of the instruction is obvious and does not need to be depicted.

This symbol indicates that gentle pressure should be applied to the model in the direction of the arrowhead.

A circle is used to identify a point on the model that is referred to in a written instruction.

This indicates that a fold is only made through the number of degrees shown.

The insert arrow is used to show how finished modules are assembled.

The ruler symbol indicates that the end of the paper should be cut off. The cut is made along the edge marked with the heavier line.

A shaded box with no outline is used to indicate that the look of the finished model will be improved if creases are not allowed to stray into the shaded area (although, of course, the axis of the fold runs all the way through).

Cube
Columbus Cubes

Cube

The way in which six modules of this type will link firmly together into a cube was first discovered by Paul Jackson of London.

Six sheets of A4 paper are required.

1

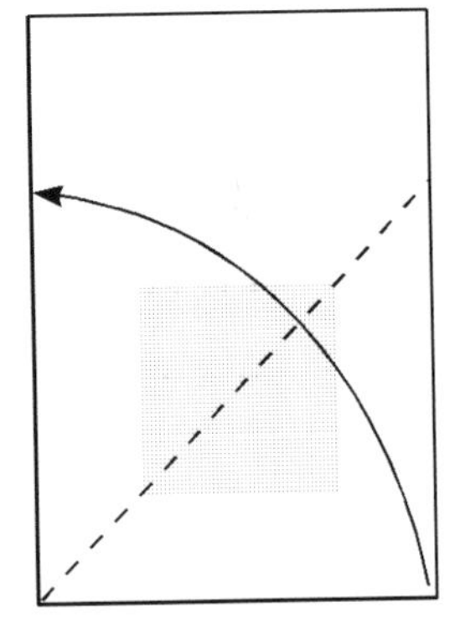

Try not to crease the paper inside the shaded area.

2

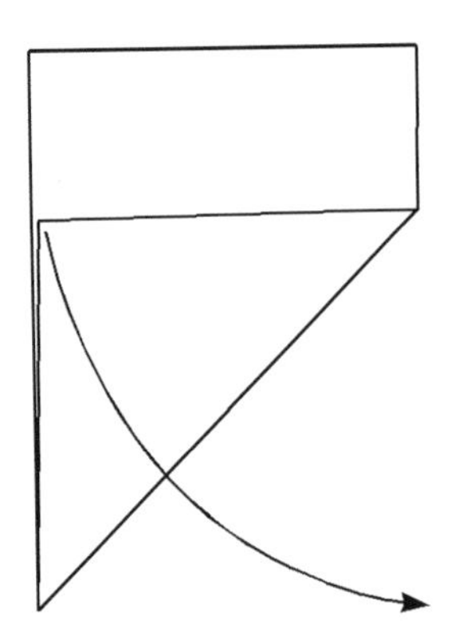

3

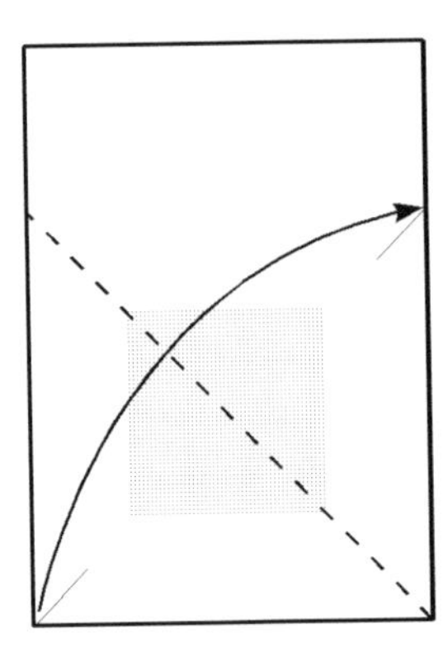

4

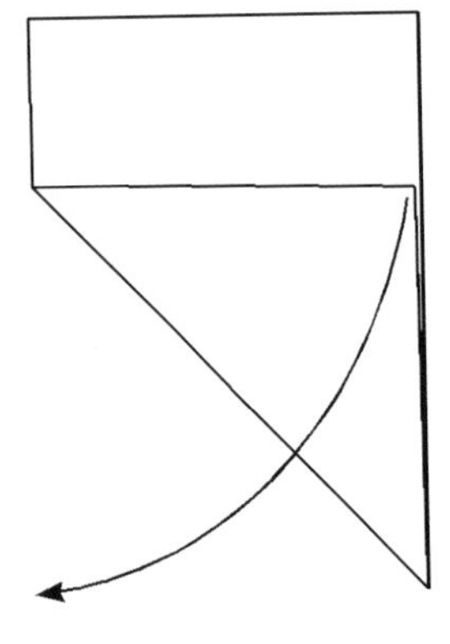

5

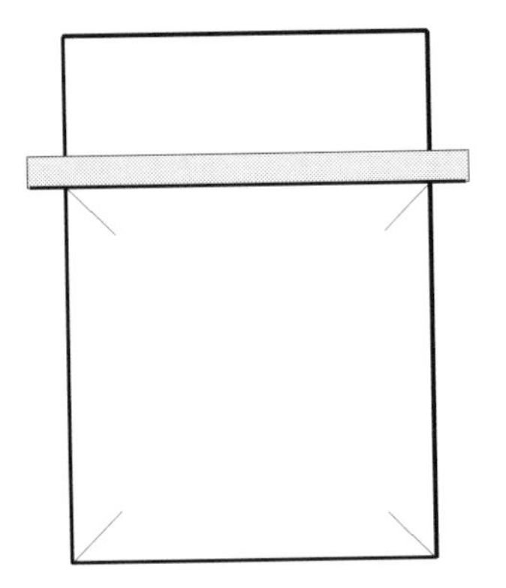

6

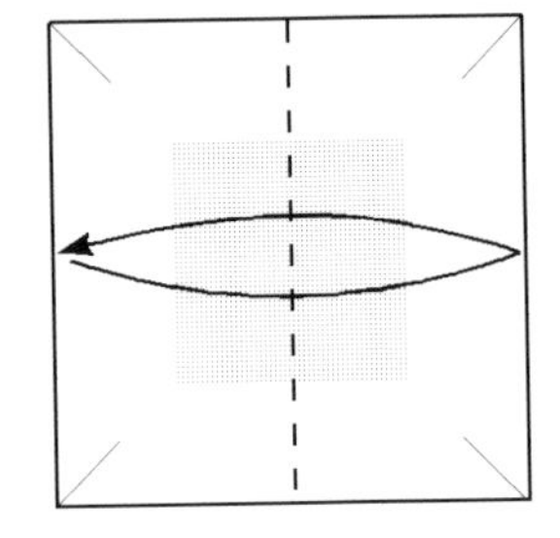

Try not to crease the paper inside the shaded area.

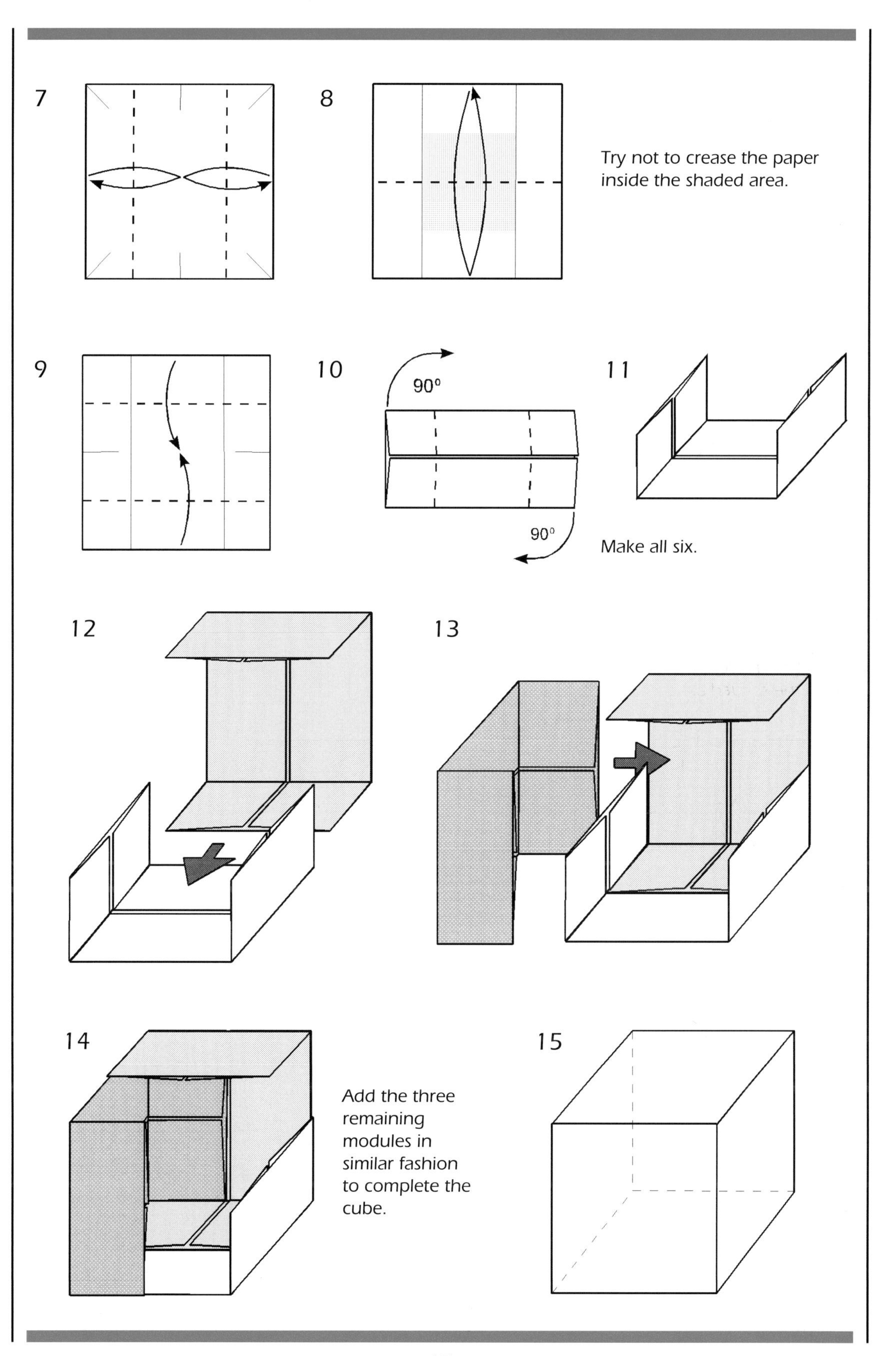
7
8
Try not to crease the paper inside the shaded area.
9
10
90°
90°
11
Make all six.
12
13
14
Add the three remaining modules in similar fashion to complete the cube.
15

Columbus Cubes

The Columbus Cube is a simple variation of the basic cube which makes it possible to combine cubes into larger structures.

Begin by taking the basic cube apart. Three of the modules must be altered as shown. When the model is assembled one corner of the cube will be inverted.

1

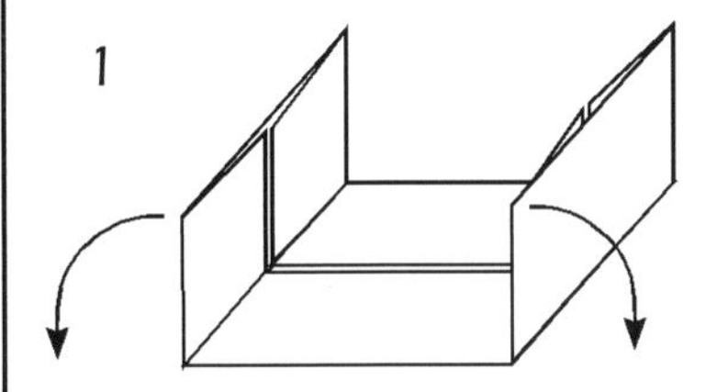

Flatten the module out.

2

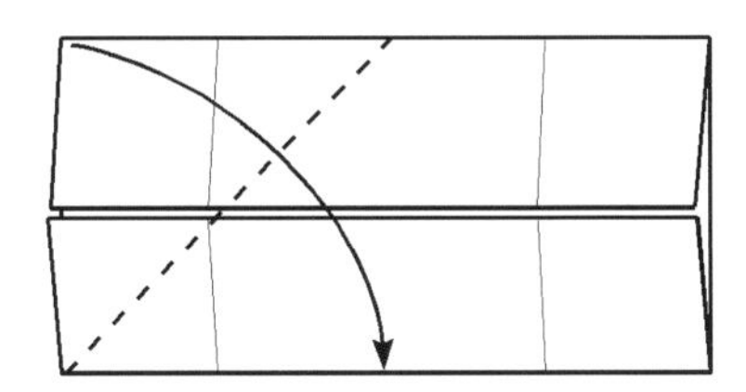

3

Make this fold, then squash the paper flat. An entirely new crease will form between the two circled points.

4

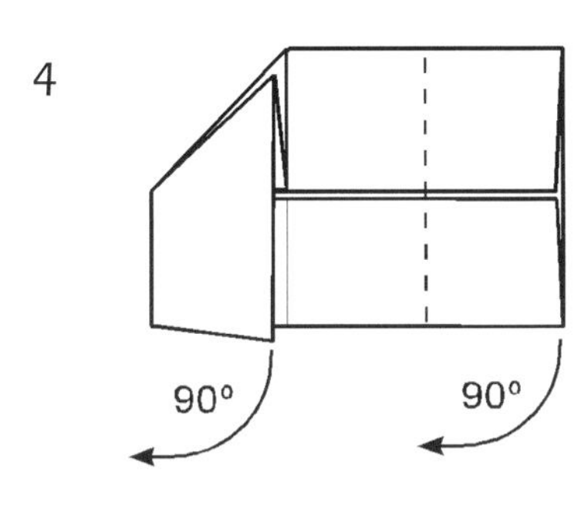

5

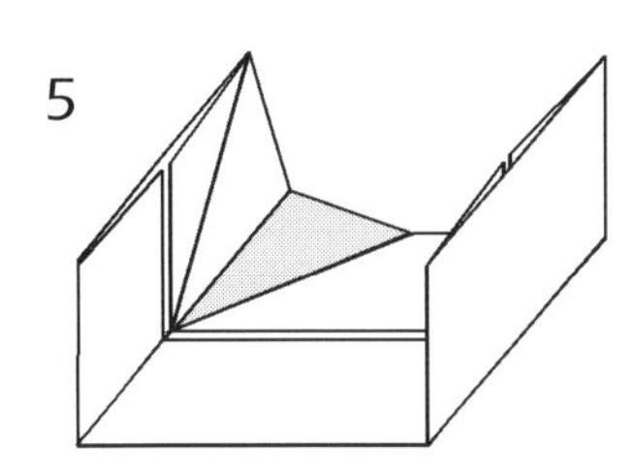

Remake three units in this way.

6

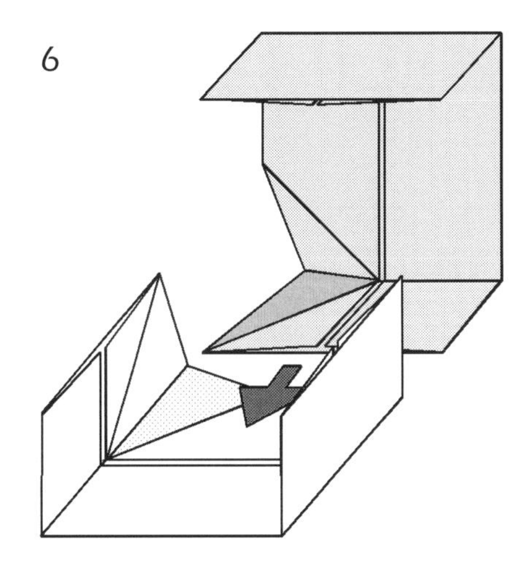

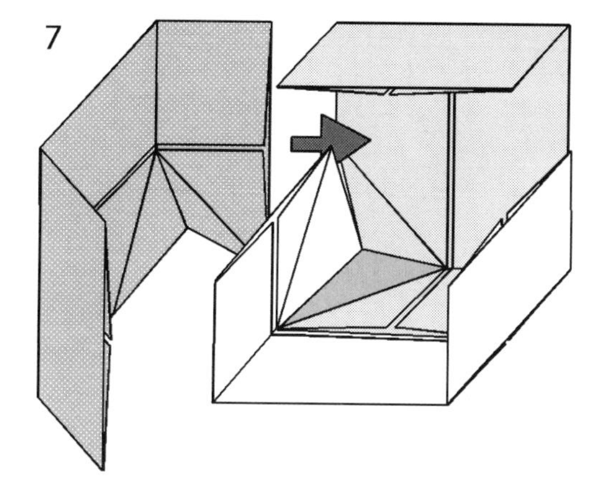

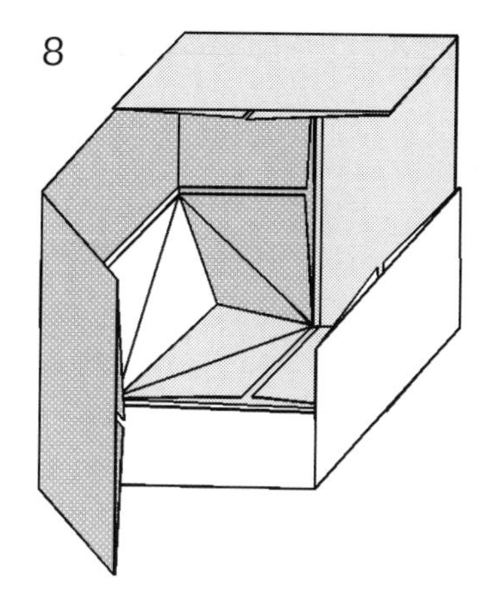

Add the other three modules to complete the Columbus Cube.

Columbus Cubes can be combined in several different ways.

Stack of Cubes

Because one corner has been pushed in, you can stand one Columbus Cube on a table so that the opposite corner points upwards, then stack as many other Columbus Cubes as you like on top. The beginning of such a stack is shown in the photograph on page 13.

If you make every face of your cubes a different colour then the colours can be made to spiral round the stack.

Ring of Cubes

If you attach the second Columbus Cube to one of the three corners adjacent to the one you have inverted, and then add three others in the same way, five Columbus Cubes will form a ring. (A little glue will be required to hold the cubes together.)

Ball of Cubes

On the same principal twenty Columbus Cubes can be combined into a dodecahedral ball consisting of twelve such rings. Each cube is shared among three rings and to complete the model you will need to invert two adjacent corners of half the cubes. This can be achieved quite simply by applying the same folding method to the modules forming the second corner.

Tetrahedron
Icosahedron
Octahedron

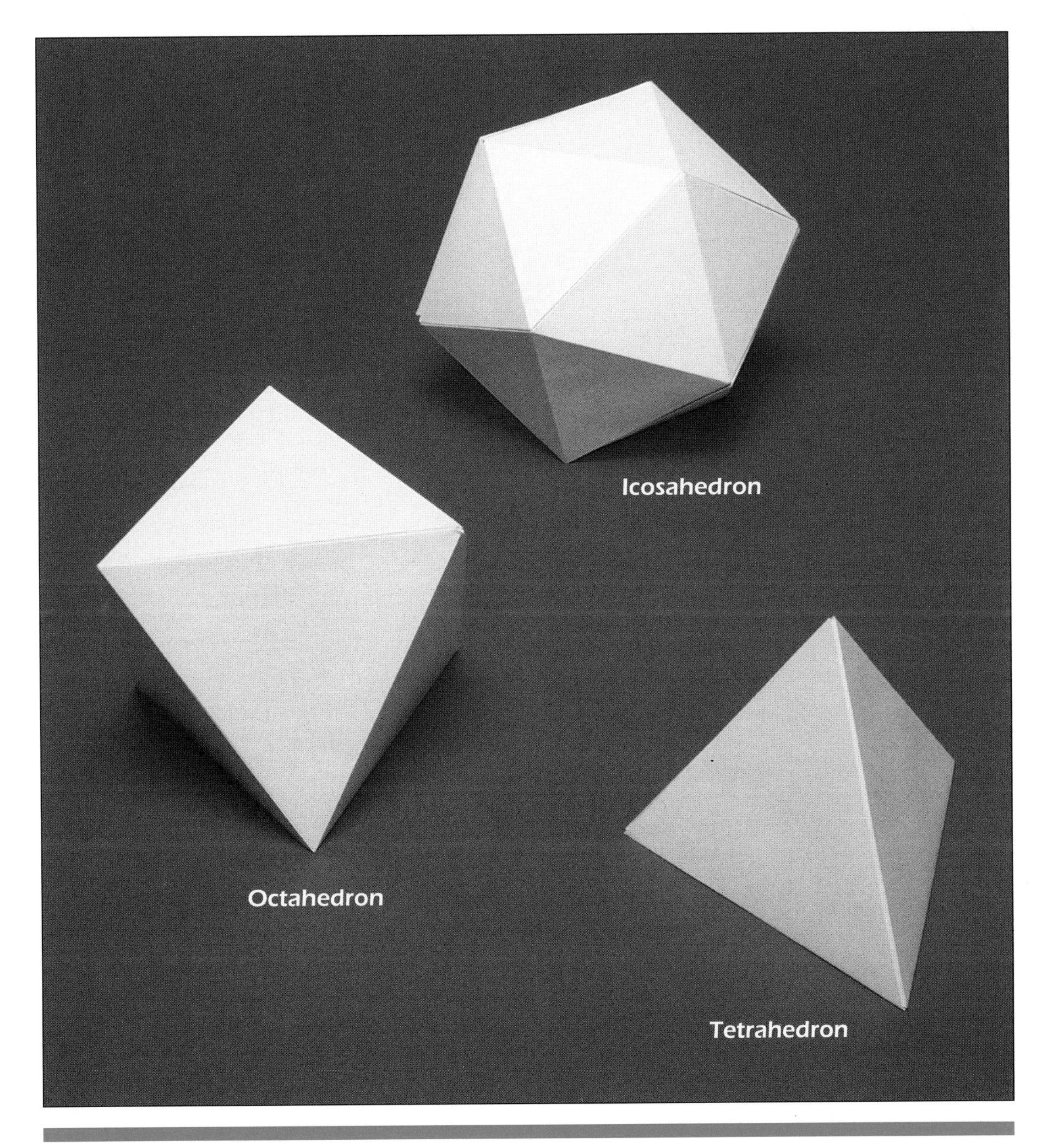

Tetrahedron

This is the first of several models in this book which use mirror image modules.

Two sheets of A4 paper are required.

The rectangle formed in step 8 is of the proportion $2:\sqrt{3}$

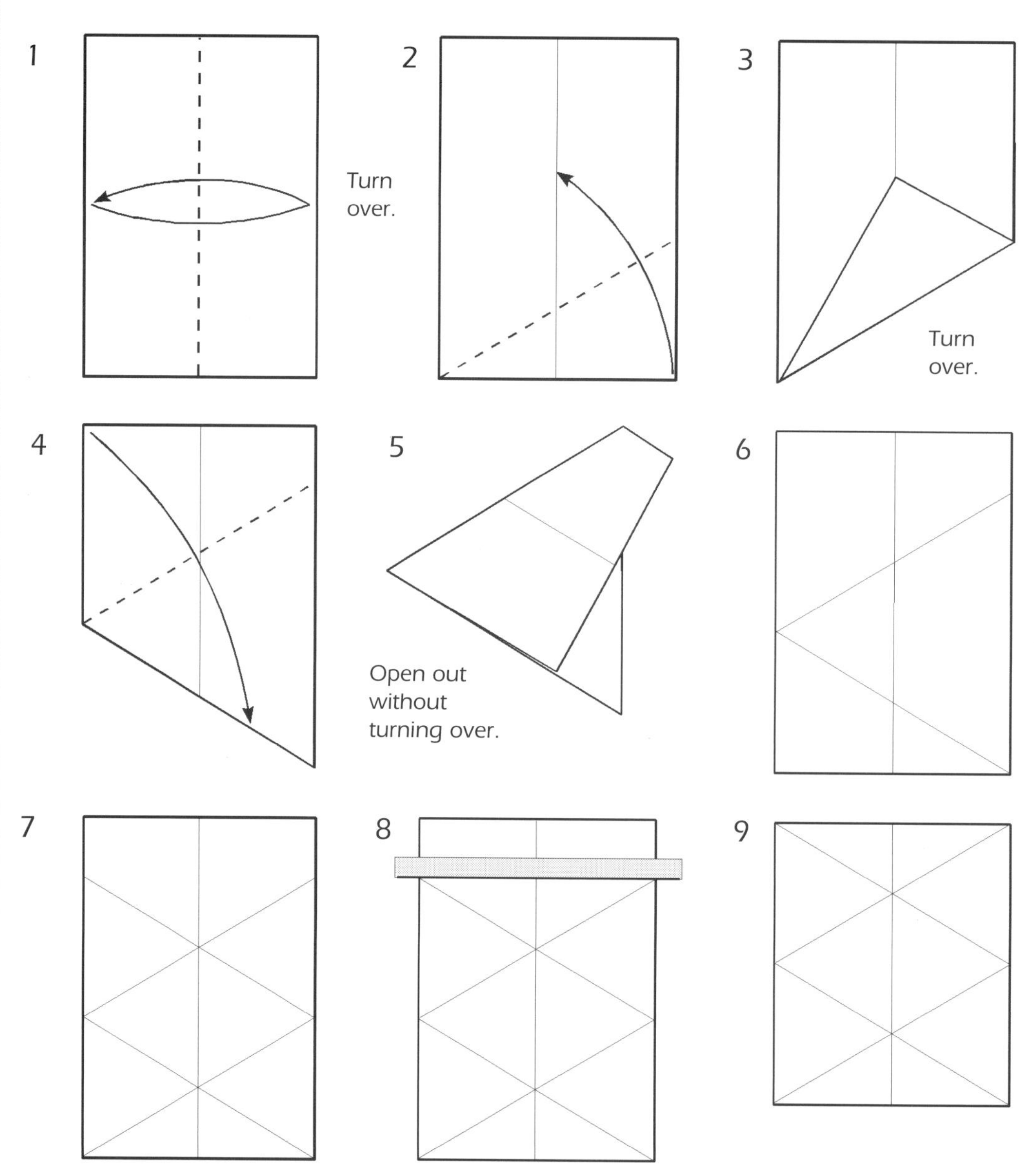

Repeat steps 2 to 5 again until the paper is creased like this.

Prepare both sheets to this point. From now on the two modules are folded as mirror images. Only the folding method for the basic module has been drawn - but if you have any difficulty making the mirror module try following the instructions for steps 10 to 13 in a mirror while you fold.

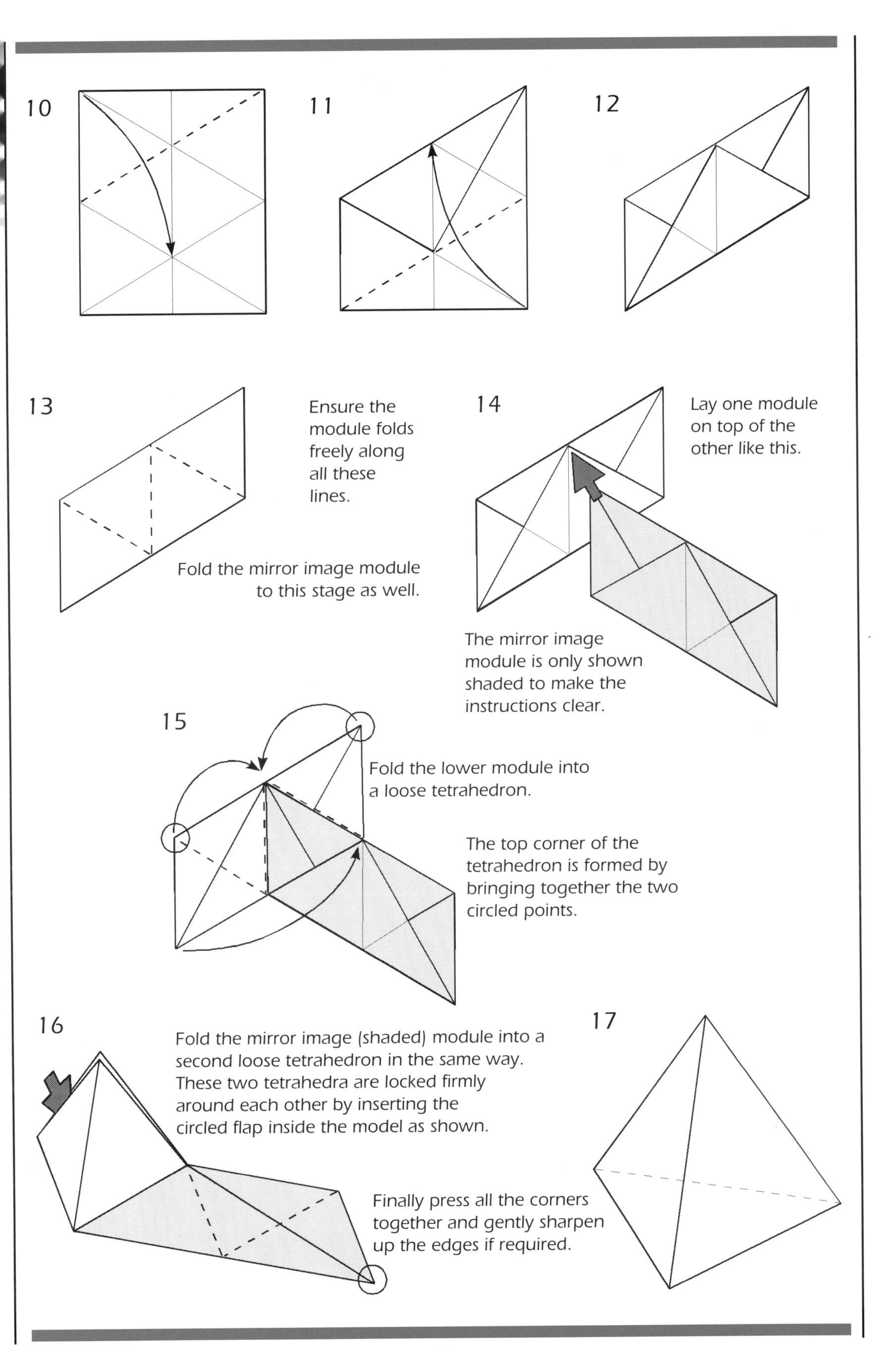
10
11
12
13
Ensure the module folds freely along all these lines.
Fold the mirror image module to this stage as well.
14
Lay one module on top of the other like this.
The mirror image module is only shown shaded to make the instructions clear.
15
Fold the lower module into a loose tetrahedron.
The top corner of the tetrahedron is formed by bringing together the two circled points.
16
Fold the mirror image (shaded) module into a second loose tetrahedron in the same way. These two tetrahedra are locked firmly around each other by inserting the circled flap inside the model as shown.
Finally press all the corners together and gently sharpen up the edges if required.
17

Icosahedron

This model was designed by Tomoko Fuse, Japan's leading lady of origami, who has published more than twenty books of modular paper-folds.

Five sheets of A4 paper are required, or even better, A3 if you can get it.

Prepare all the sheets to step 9 of the Tetrahedron.

1

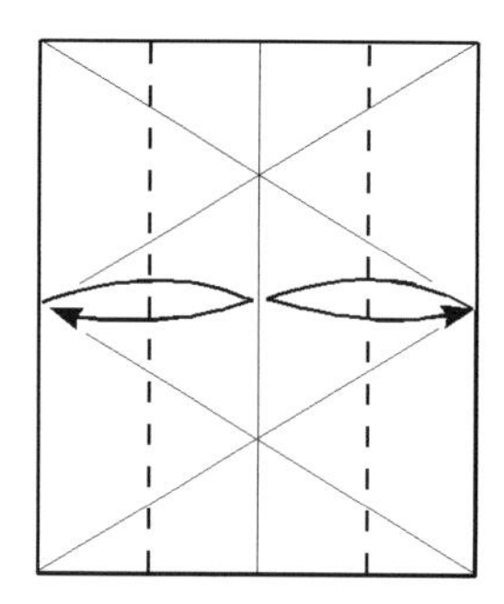

2

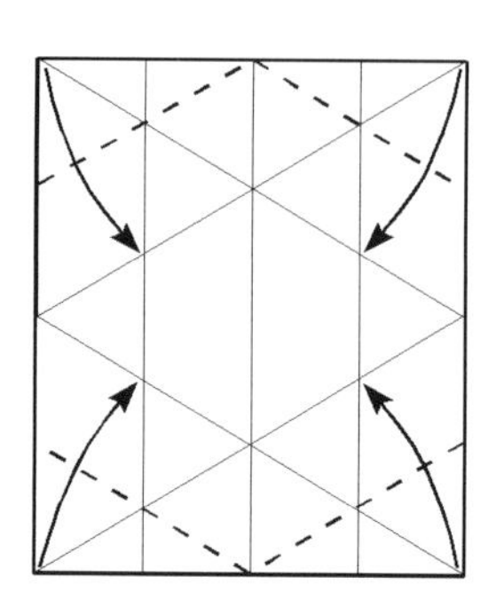

3

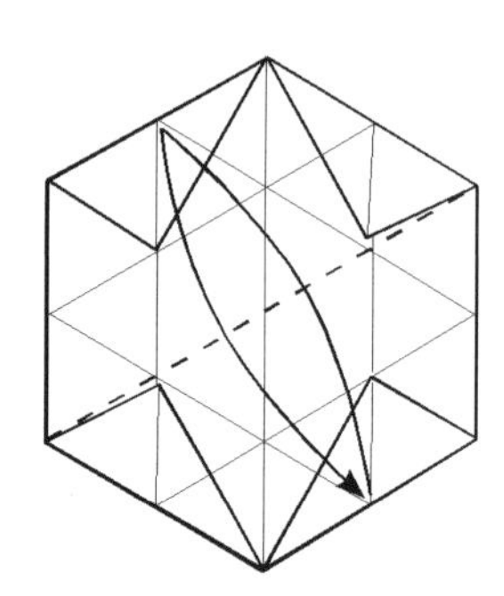

4

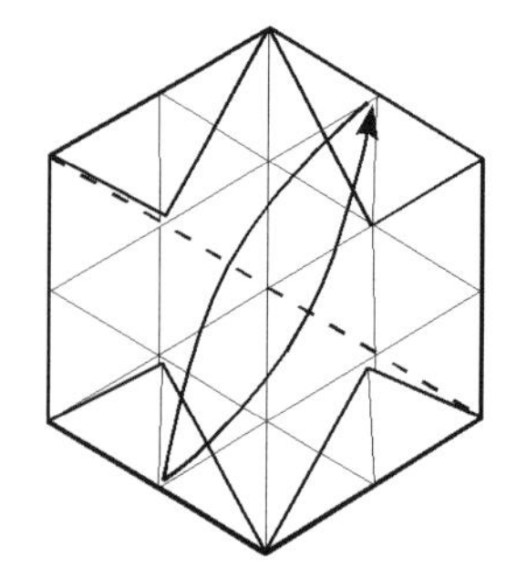

5

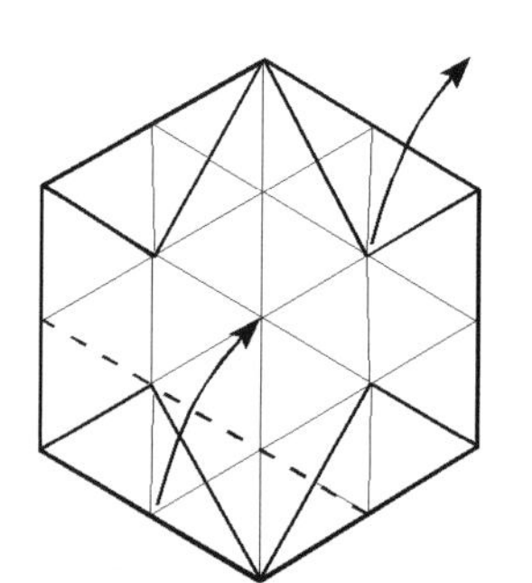

6

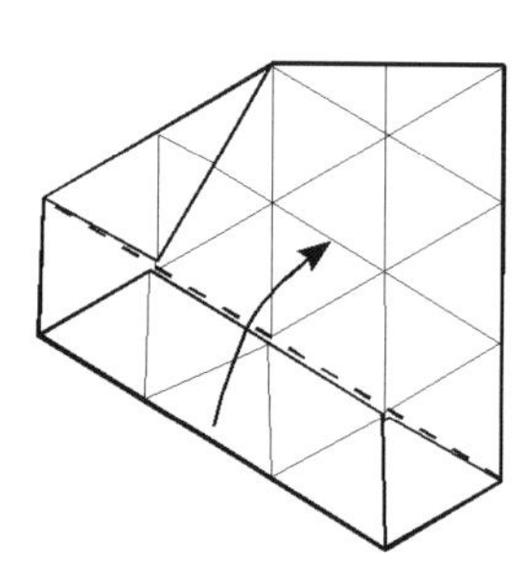

7

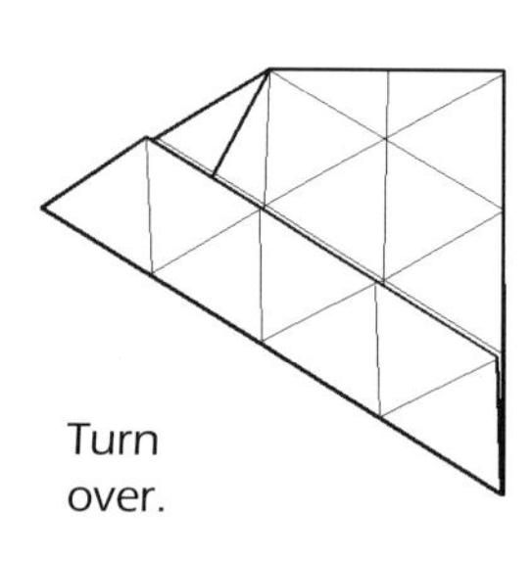

8

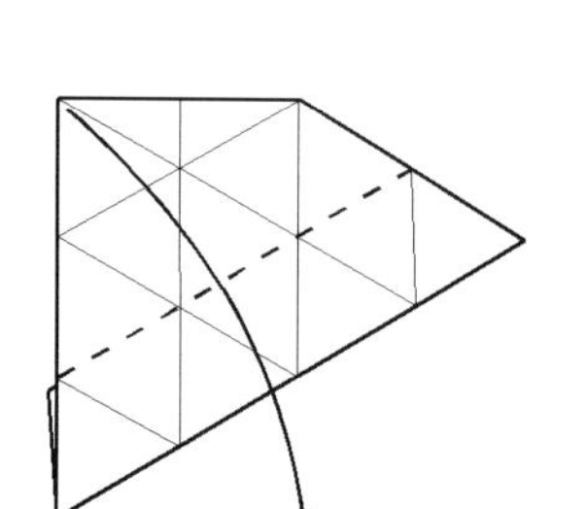

9

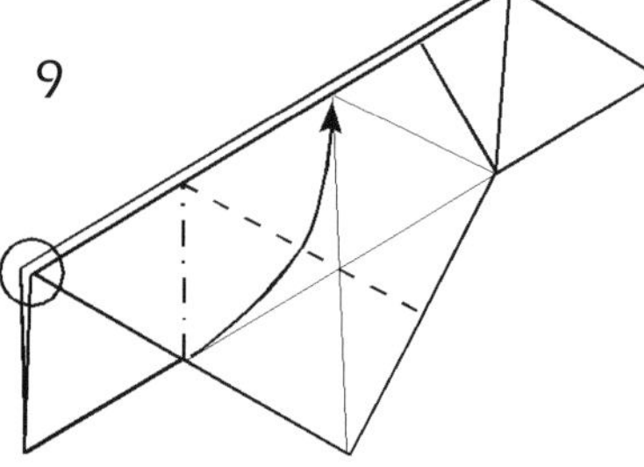

As you make this fold the flaps at the point circled will open out. Squash them down in their new position without adding any extra creases.

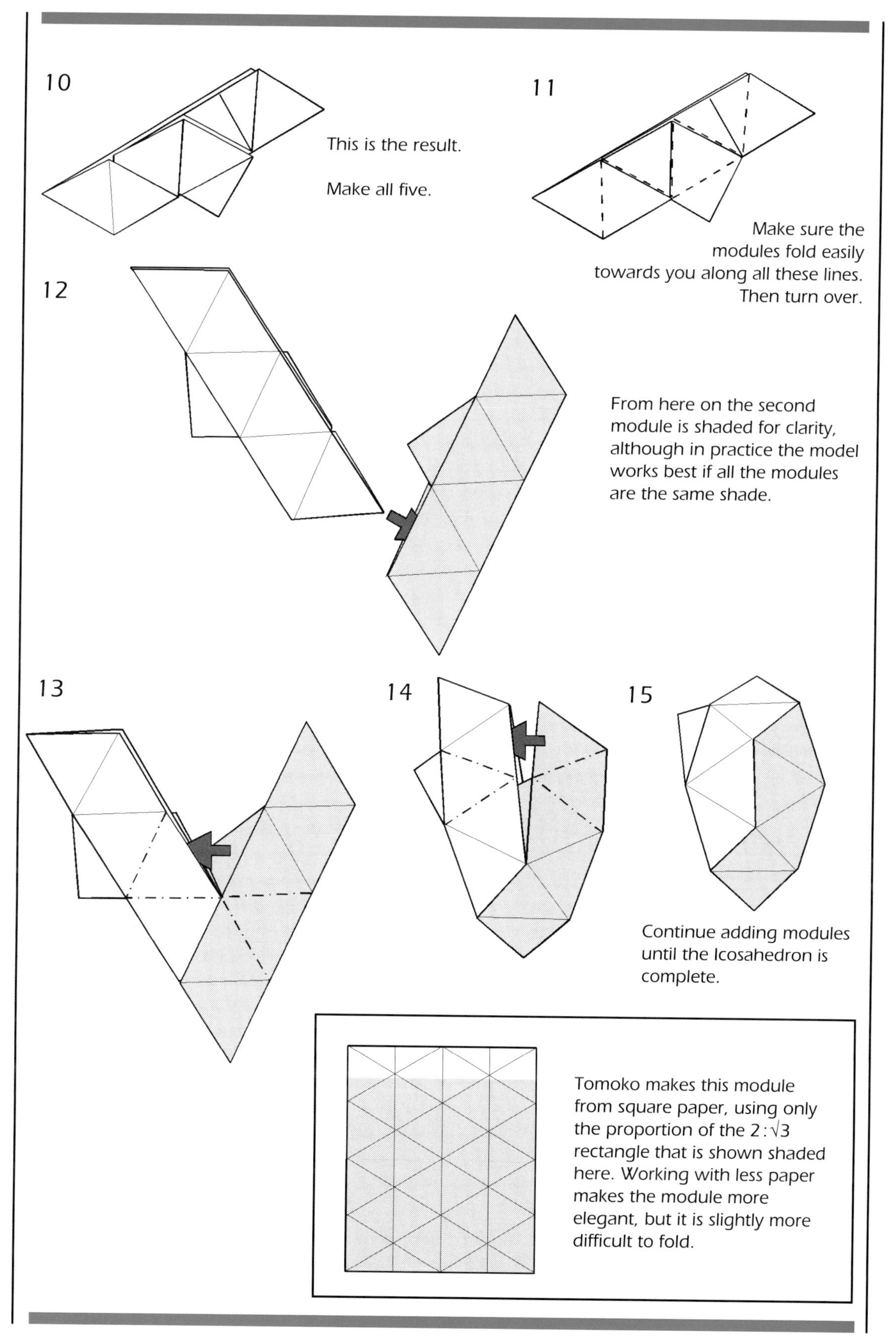
10
This is the result.
Make all five.
11
Make sure the
modules fold easily
towards you along all these lines.
Then turn over.
12
From here on the second
module is shaded for clarity,
although in practice the model
works best if all the modules
are the same shade.
13
14
15
Continue adding modules
until the Icosahedron is
complete.
Tomoko makes this module
from square paper, using only
the proportion of the 2:√3
rectangle that is shown shaded
here. Working with less paper
makes the module more
elegant, but it is slightly more
difficult to fold.

Octahedron

Two sheets of A4 paper are required.

Begin by preparing both sheets to step 9 of the Tetrahedron.

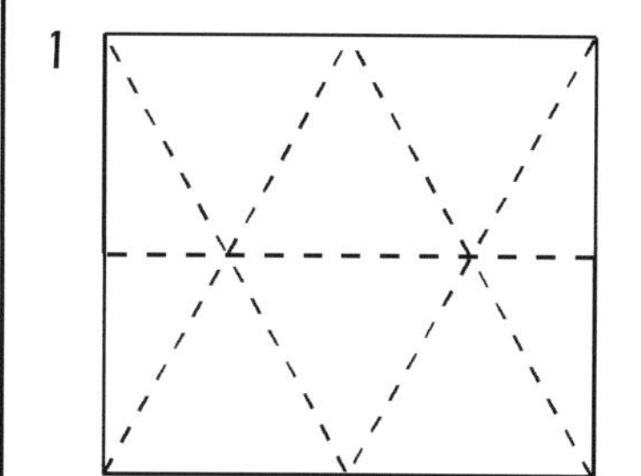

Make sure all the creases fold easily towards you.
Turn over.

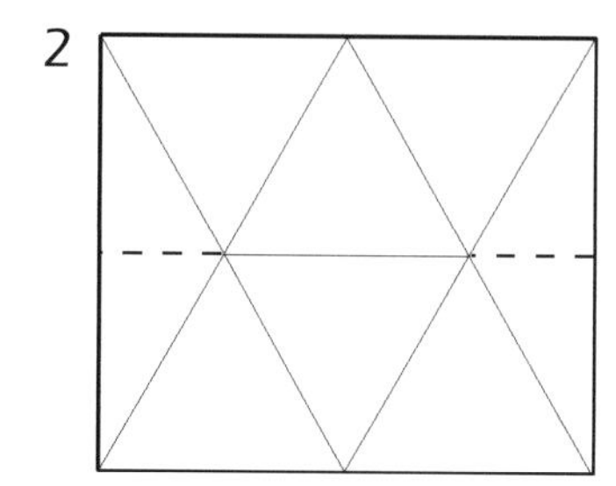

Reverse the direction of these two small creases, but make sure you don't reverse any part of the crease between them.

Turn back over.

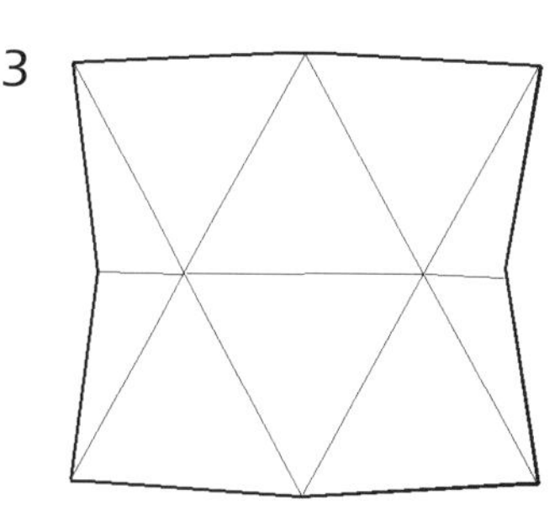

Allow the paper to remain slightly three dimensional.

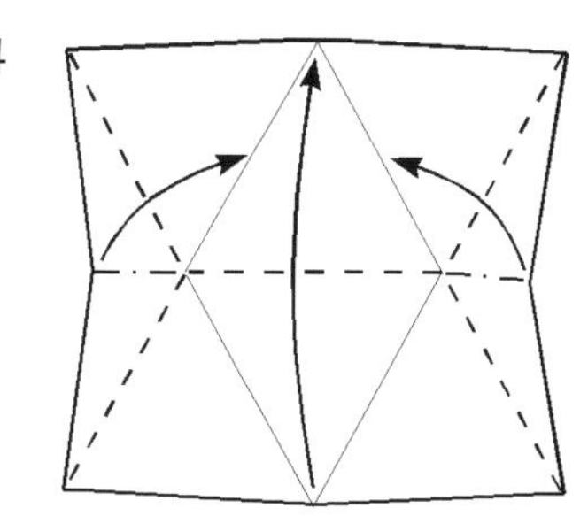

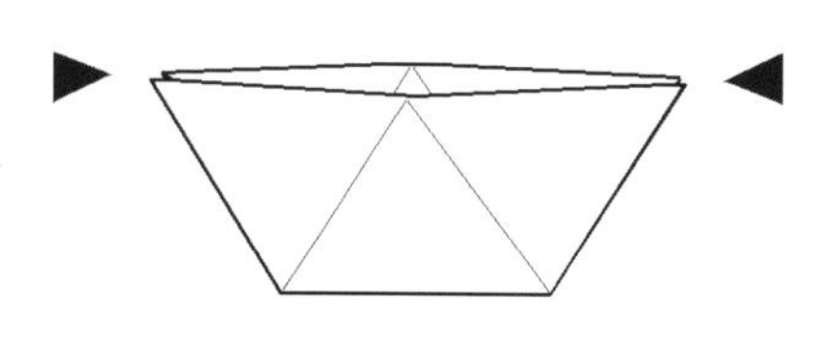

Apply gentle pressure to the module as shown and it will become three–dimensional like this.

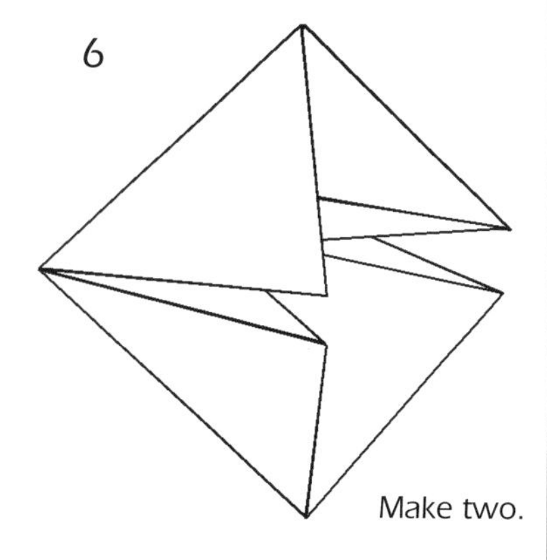

Make two.

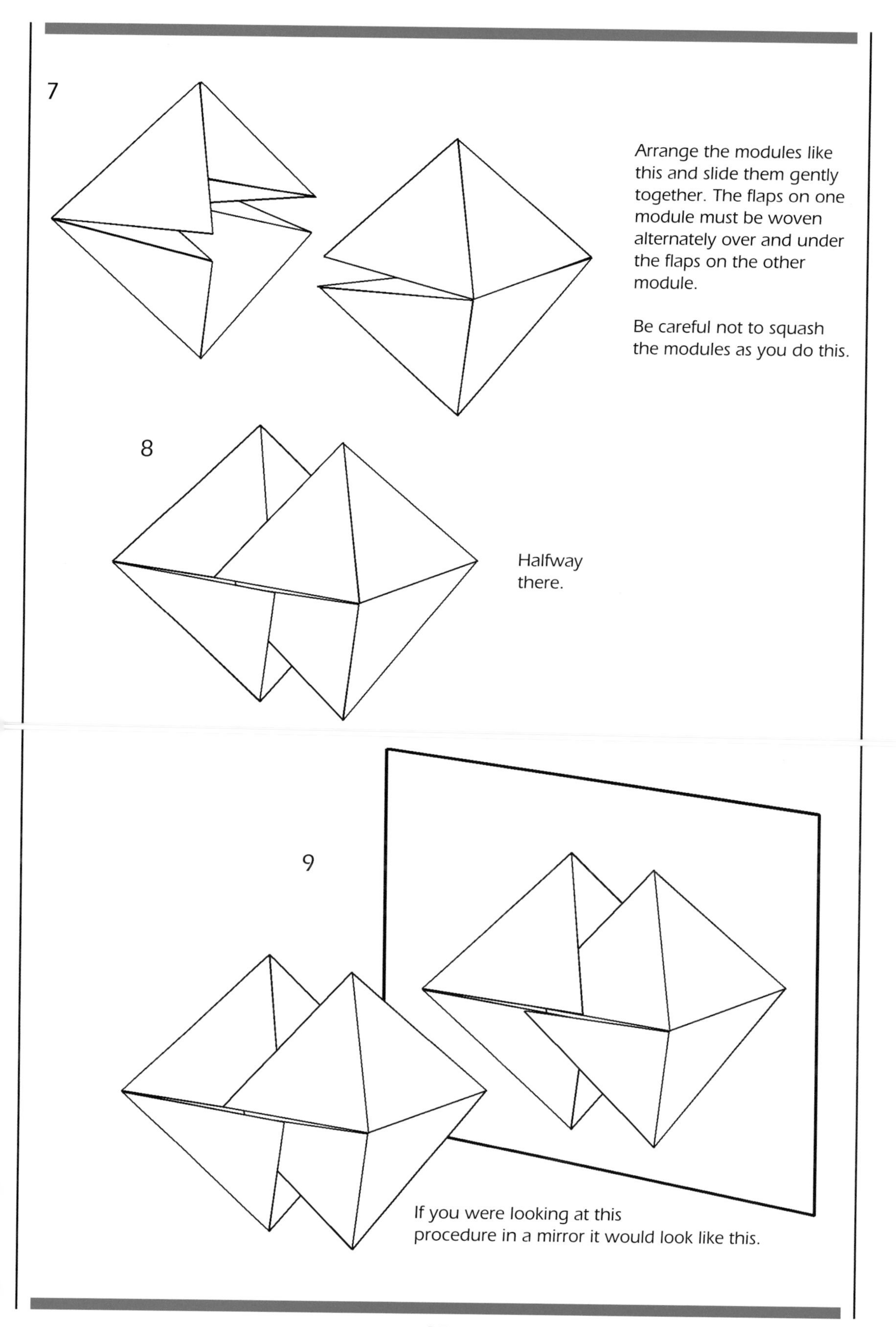
7
Arrange the modules like this and slide them gently together. The flaps on one module must be woven alternately over and under the flaps on the other module.
Be careful not to squash the modules as you do this.
8
Halfway there.
9
If you were looking at this procedure in a mirror it would look like this.

Because the faces of the regular Tetrahedron and Octahedron are to scale - when made from the same size paper - they can be combined to pack space in various ways and to produce the familiar Compound Tetrahedron like this:

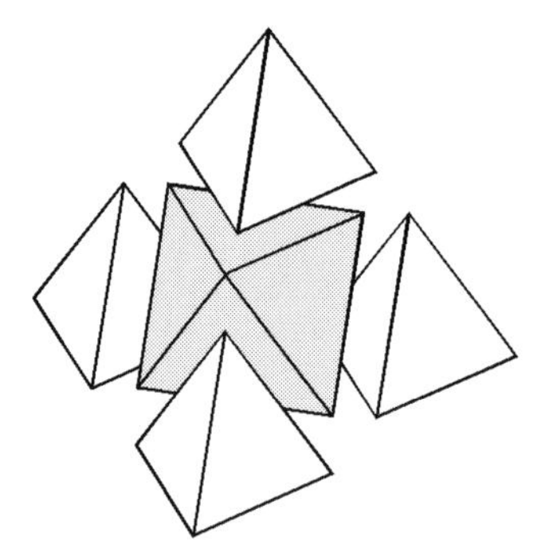

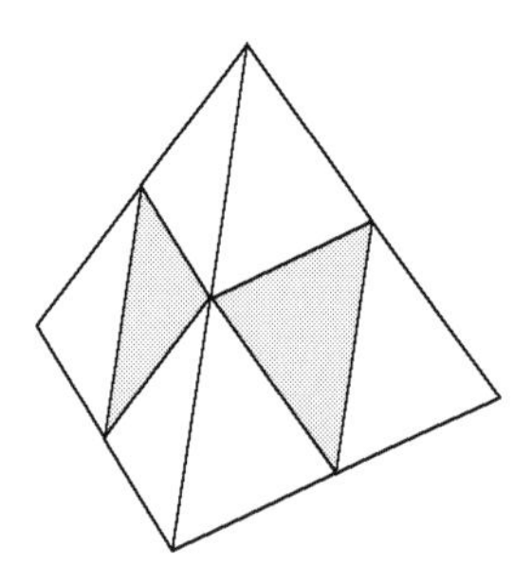

Skeletal Octahedron
Skeletal Cuboctahedron
Skeletal Cube

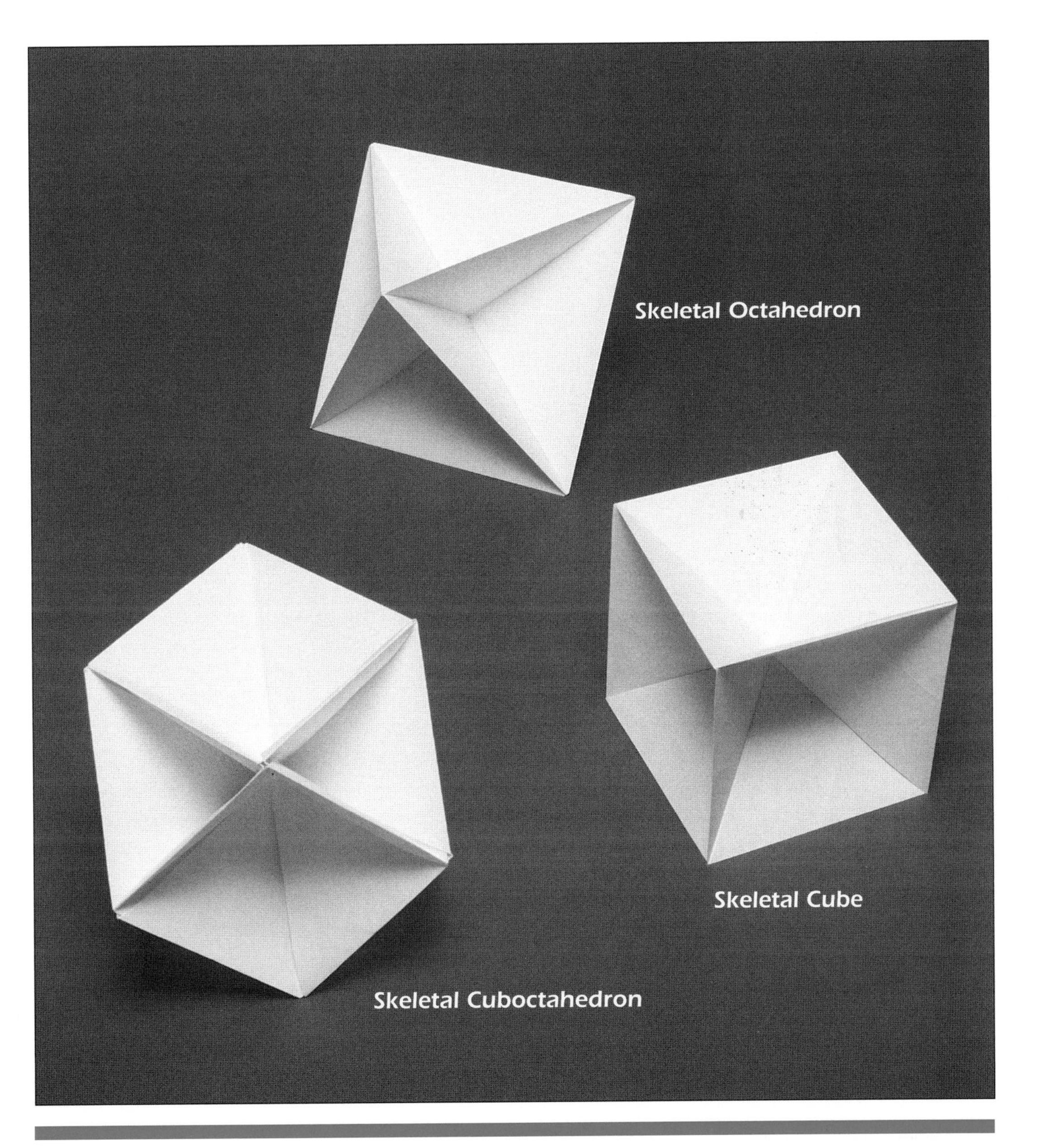

Skeletal Octahedron

The discovery of this model by Bob Neale of the USA in the early 1960's was the inspirational seed from which much of the best of modular origami has since grown. It demonstrated that modular origami is not only an effective technique for modelling polyhedra but can also, at its best, be a tool for the investigation and development of shape and form.

Six sheets of A4 paper are required. To highlight the intersecting planes use two sheets in each of three contrasting colours.

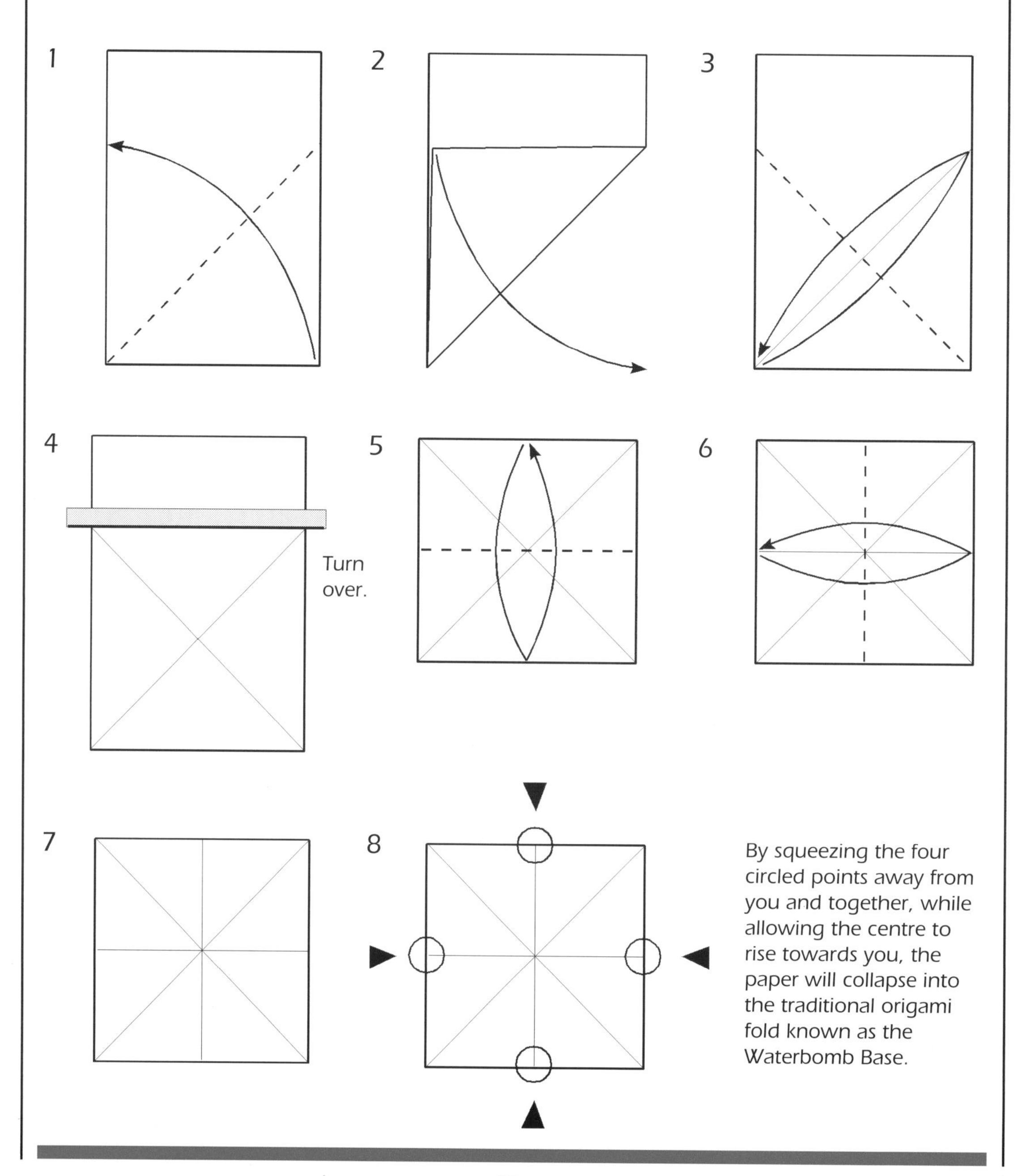

By squeezing the four circled points away from you and together, while allowing the centre to rise towards you, the paper will collapse into the traditional origami fold known as the Waterbomb Base.

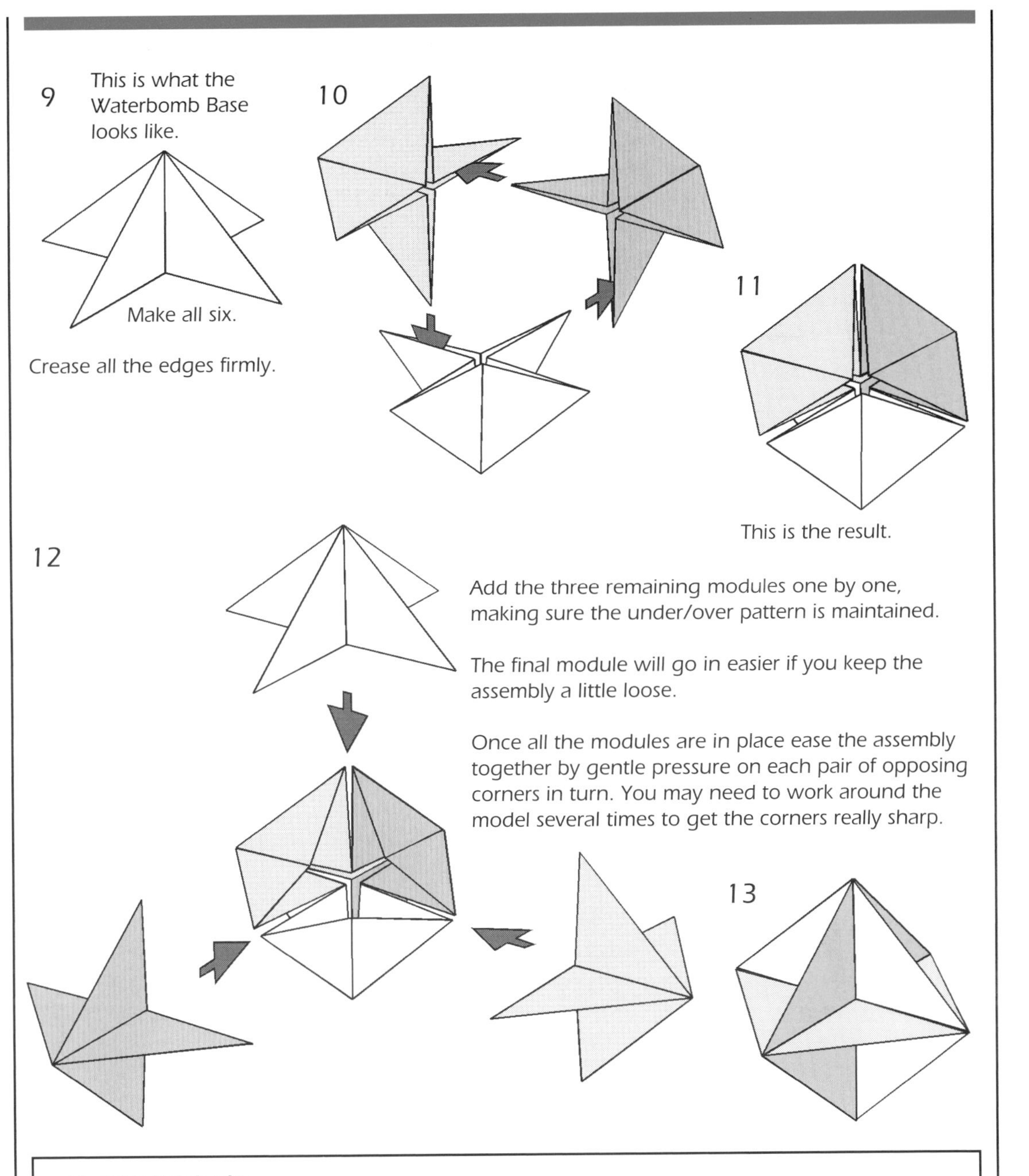

Skeletal Polyhedra

Sometimes known as Nolids, these interesting forms are composed of finite planes which stretch between each edge and the centre of symmetry of the polyhedron. The next model - the Skeletal Cube - demonstrates a general method for producing modules for many skeletal polyhedra. All that is required is a rectangle whose diagonals intersect at the correct angle. A Golden Rectangle will, for instance, produce the correct central angle to model a Skeletal Icosahedron.

There are three special cases in which the planes combine to form interpenetrating polygons. The Skeletal Octahedron consists of three interpenetrating squares, the Skeletal Cuboctahedron of four interpenetrating hexagons and the Skeletal Icosidodecahedron (unfortunately beyond the scope of this book) of six interpenetrating decagons.

Skeletal Cuboctahedron

Twelve sheets of A4 paper are required. To highlight the intersecting hexagonal planes use three sheets in each of four contrasting colours.

Begin by halving each A4 sheet to A5 and preparing all the A5 sheets to step 9 of the Tetrahedron.

1

2

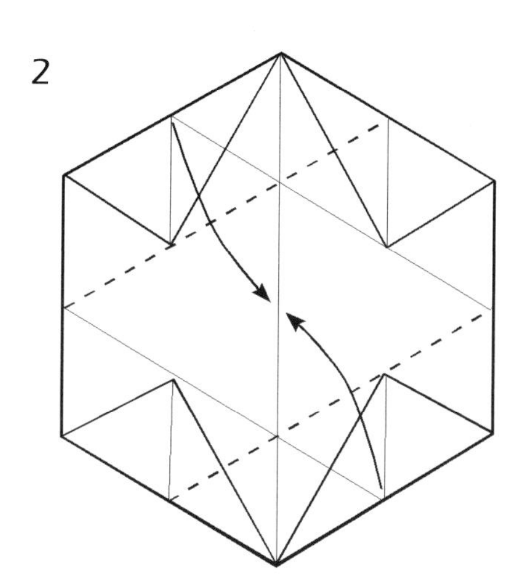

3

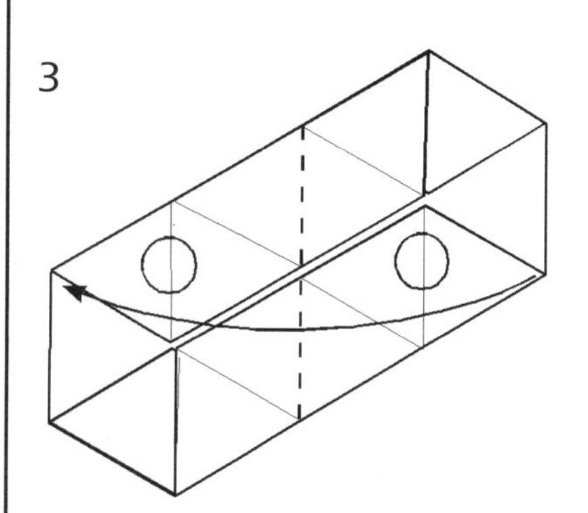

As you make this fold interlock the two flaps marked with a circle to hold the two sides of the module together.

4

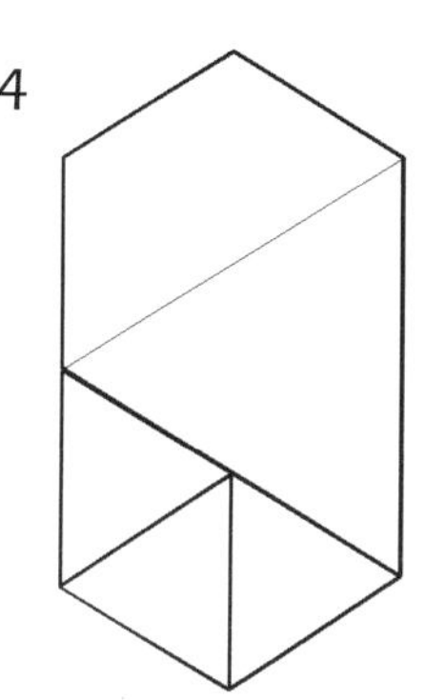

5

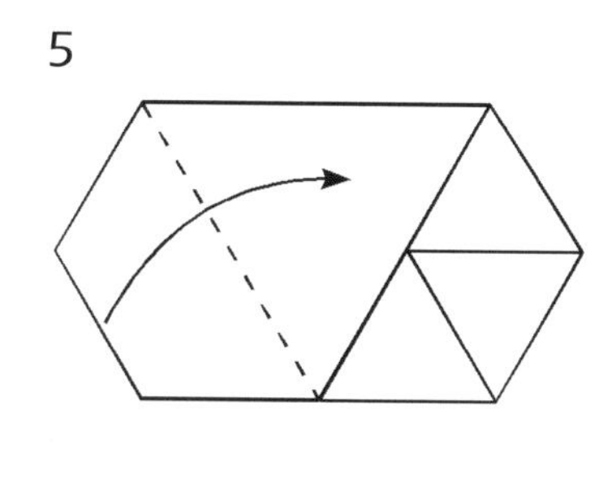

6

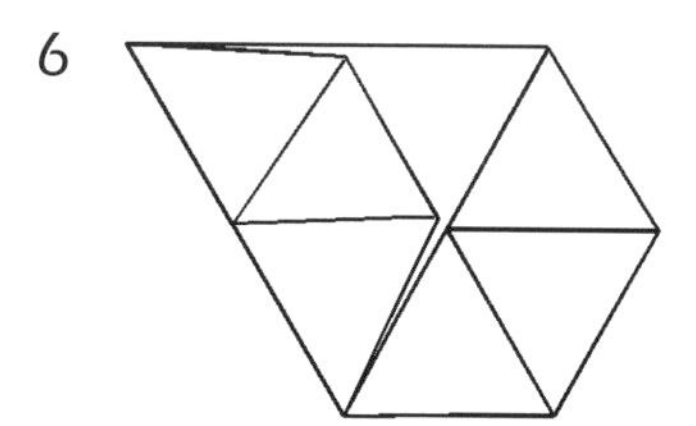

Turn over.

7

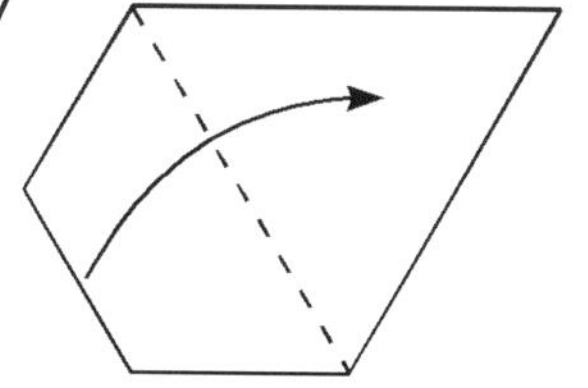

Crease firmly.

8

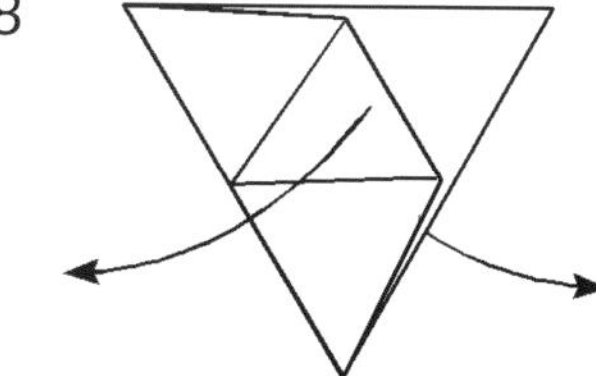

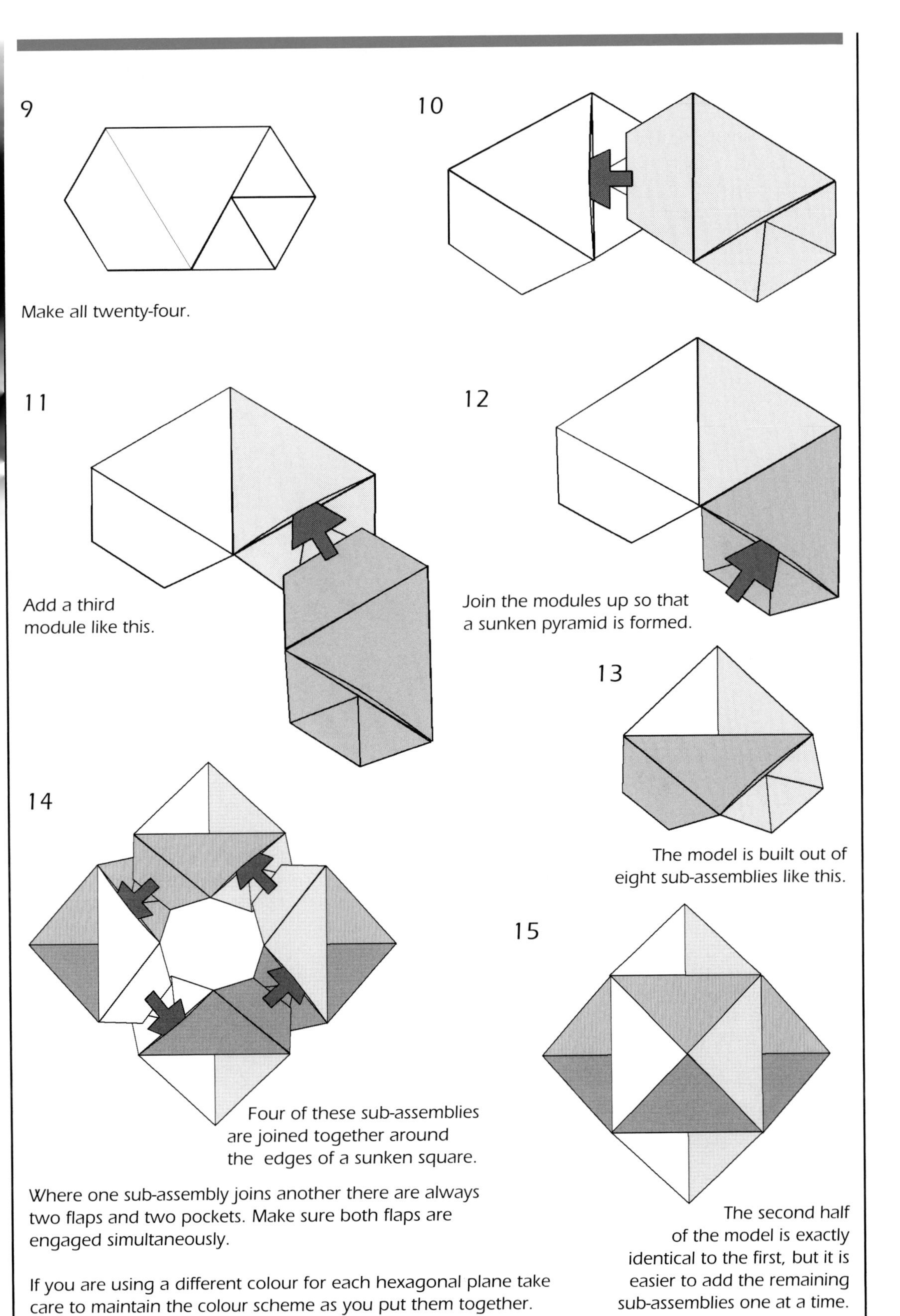

9

Make all twenty-four.

10

11

Add a third module like this.

12

Join the modules up so that a sunken pyramid is formed.

13

The model is built out of eight sub-assemblies like this.

14

Four of these sub-assemblies are joined together around the edges of a sunken square.

Where one sub-assembly joins another there are always two flaps and two pockets. Make sure both flaps are engaged simultaneously.

If you are using a different colour for each hexagonal plane take care to maintain the colour scheme as you put them together.

15

The second half of the model is exactly identical to the first, but it is easier to add the remaining sub-assemblies one at a time.

Skeletal Cube

This wonderfully simple model was designed by David Brill of Stockport.

Three sheets of A4 paper are required. Divide each sheet into quarters in the way shown before you start to fold the modules.

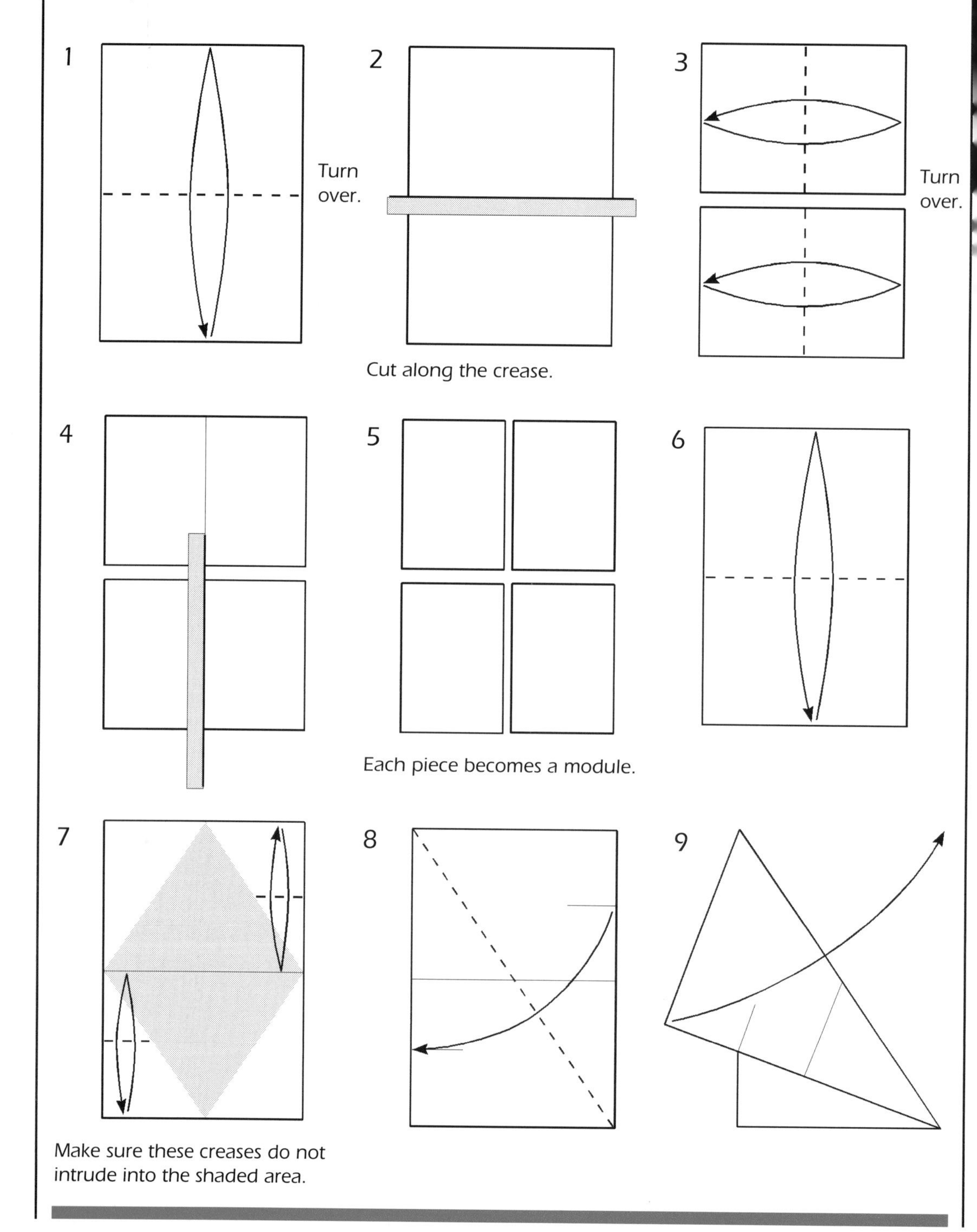

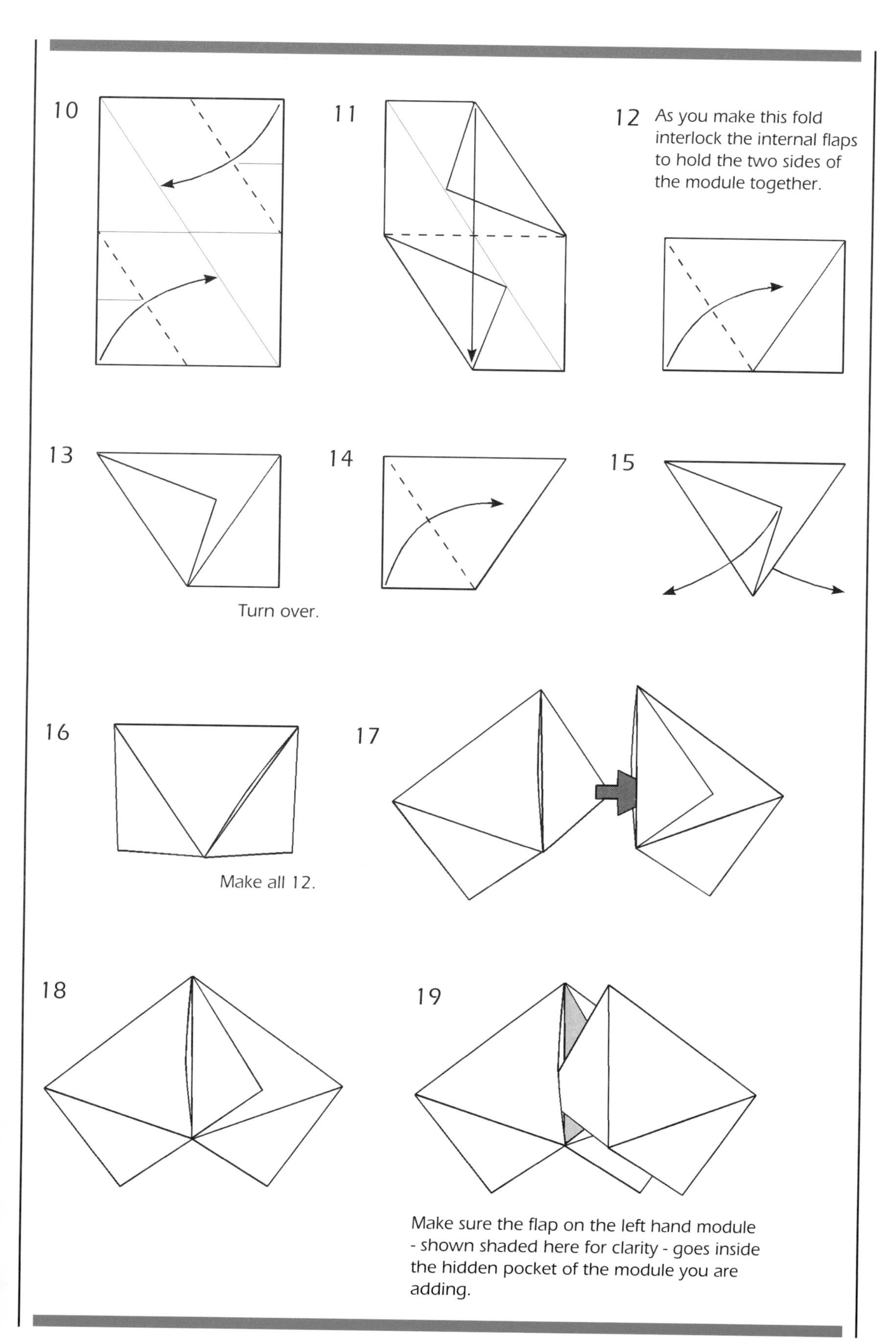
10
11
12
As you make this fold interlock the internal flaps to hold the two sides of the module together.
13
Turn over.
14
15
16
Make all 12.
17
18
19
Make sure the flap on the left hand module - shown shaded here for clarity - goes inside the hidden pocket of the module you are adding.

20

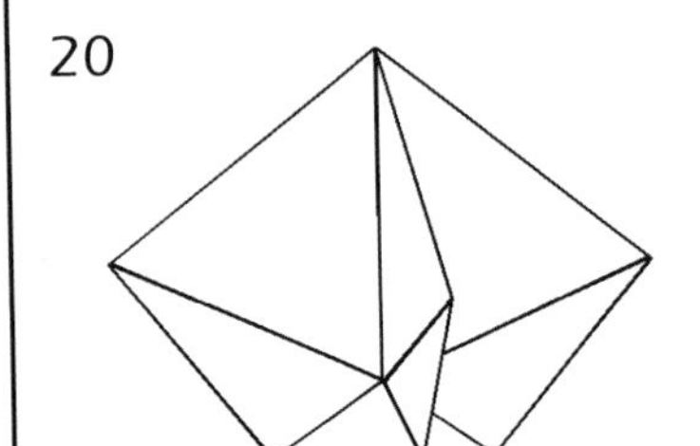

Make four of these.

21

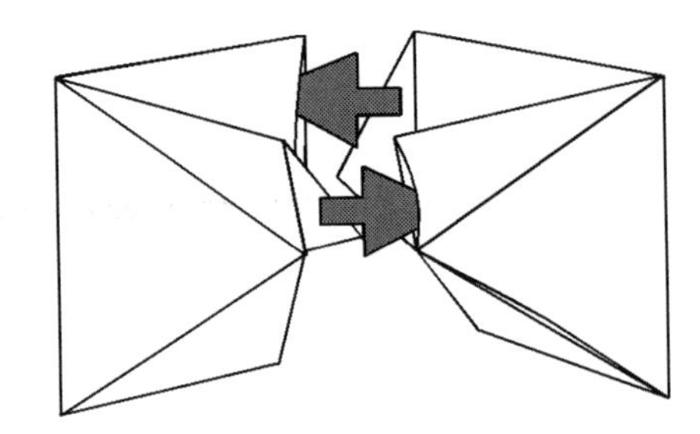

Join two modules.

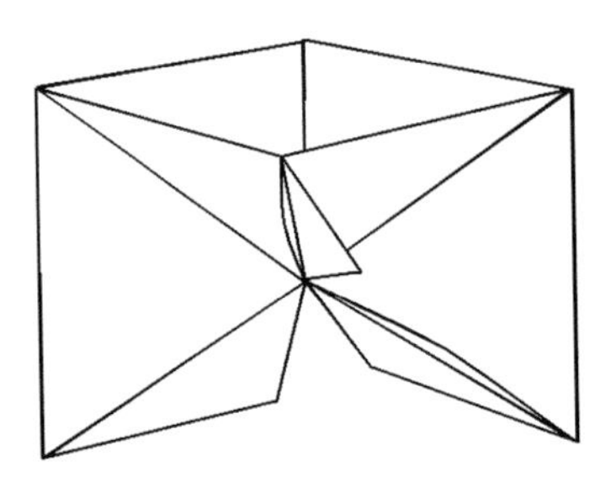

Add the remaining modules to complete the Skeletal Cube.

Rhombic Dodecahedron
Rhombic Pyramid
Rhombic Tetrahedron

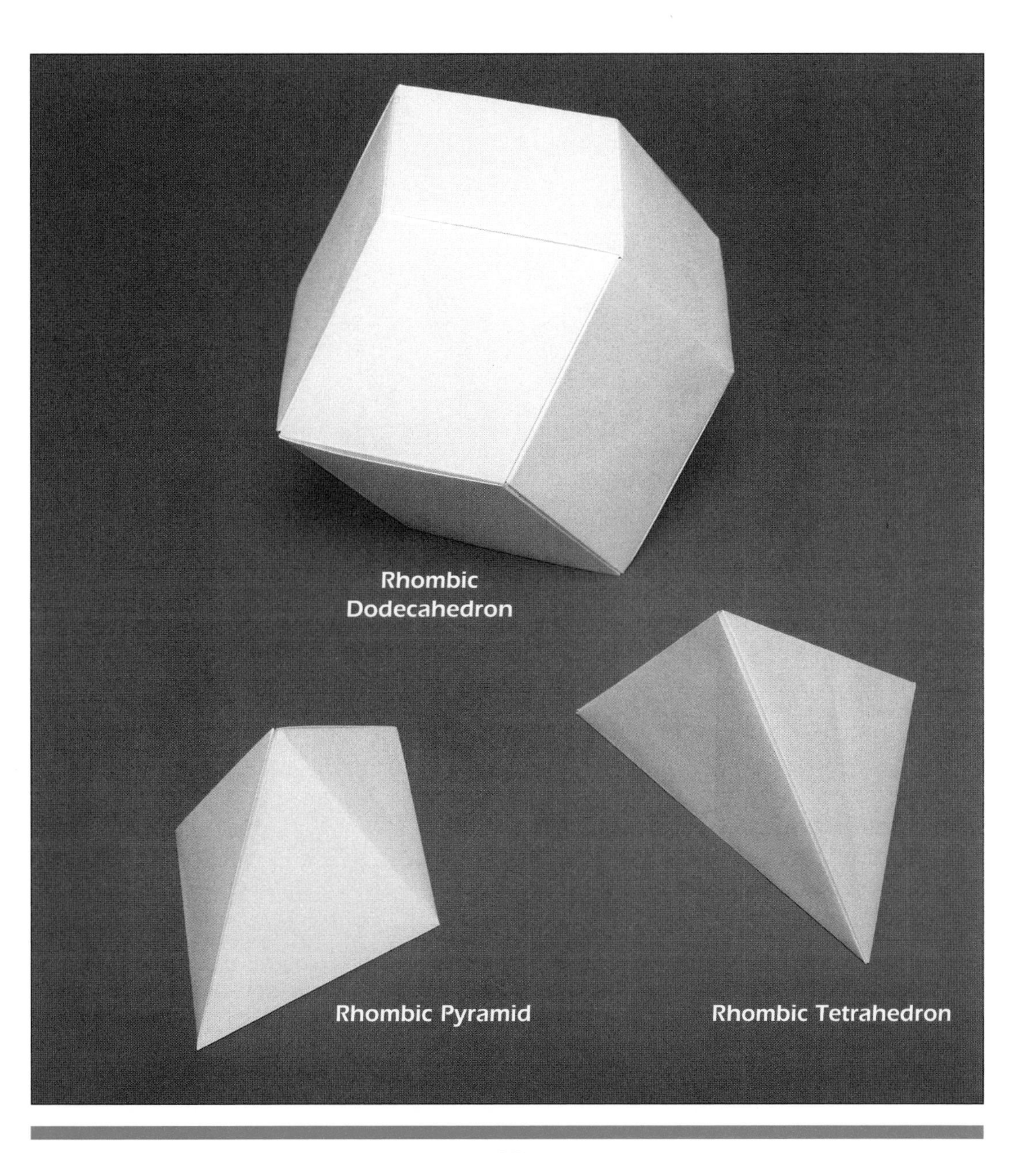

Rhombic Dodecahedron

This classic model was designed by Nick Robinson of Sheffield.

Twelve sheets of A4 paper are required.

1

Try not to let your creases stray into the shaded area.

2

Turn over

3

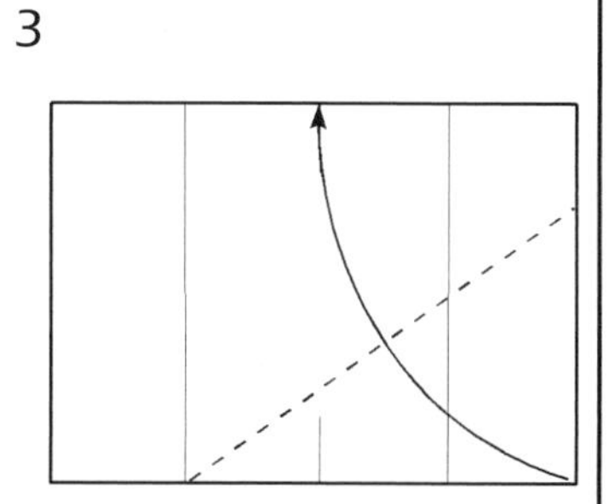

4

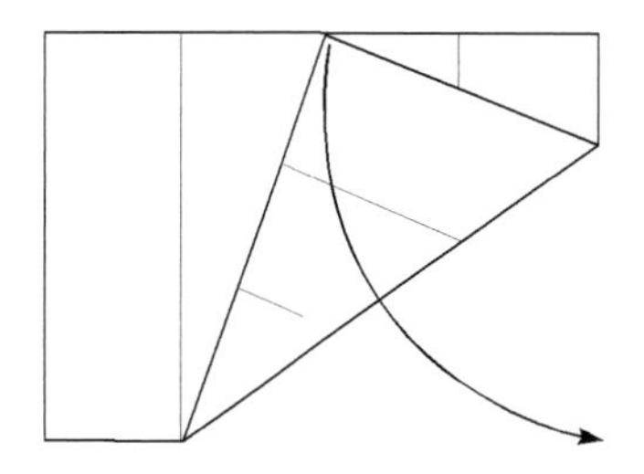

5

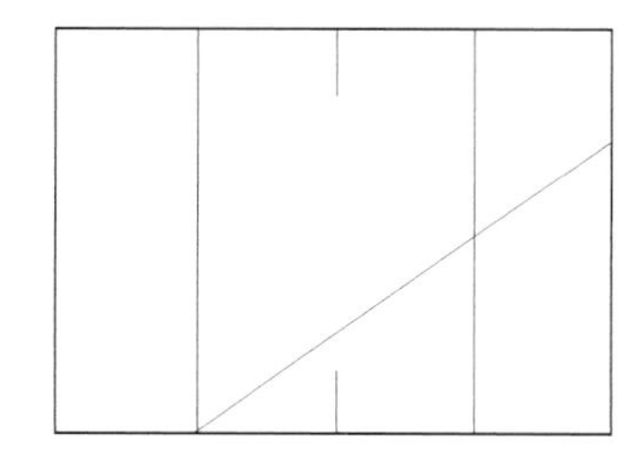

Repeat folds 3 and 4 on all the other corners.

6

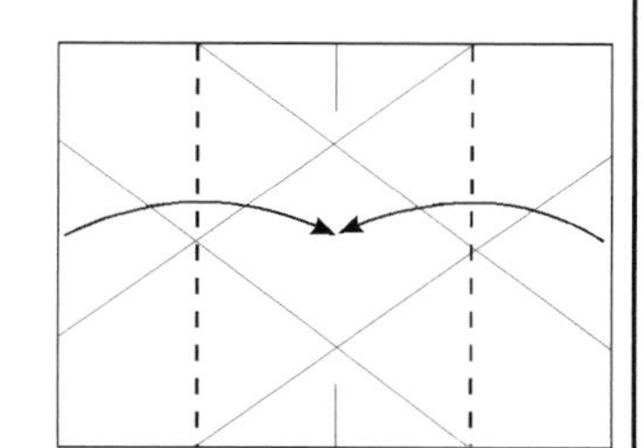

7

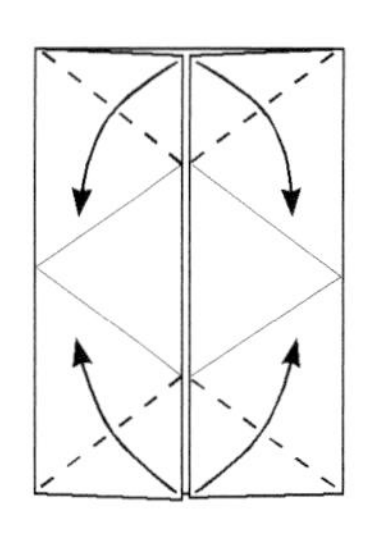

8

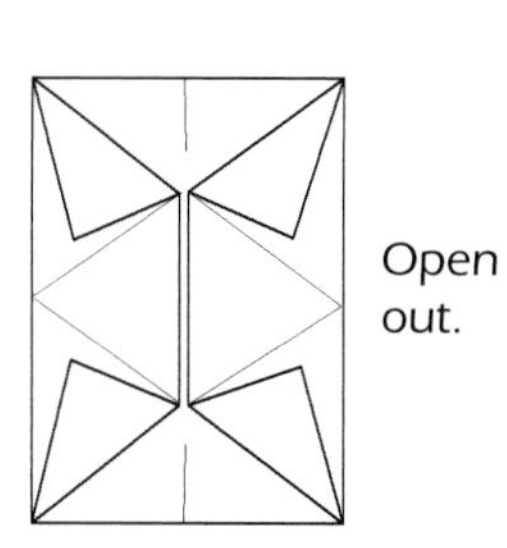

Open out.

9

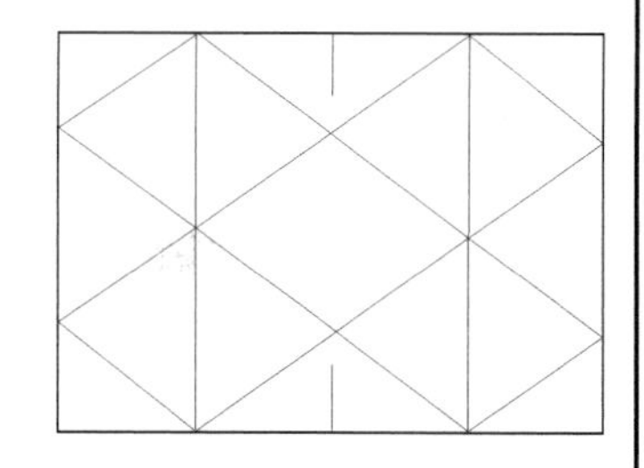

10

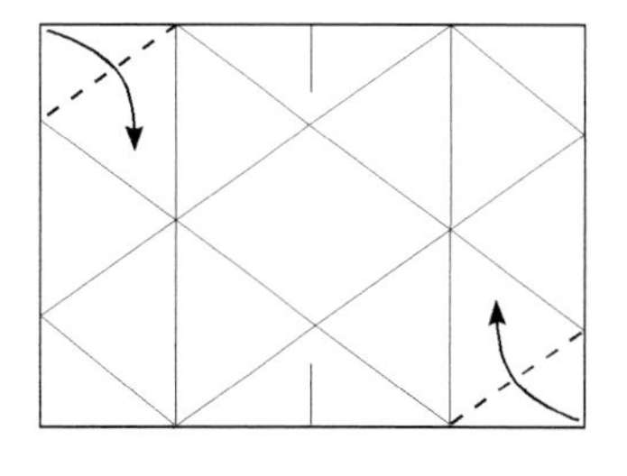

11

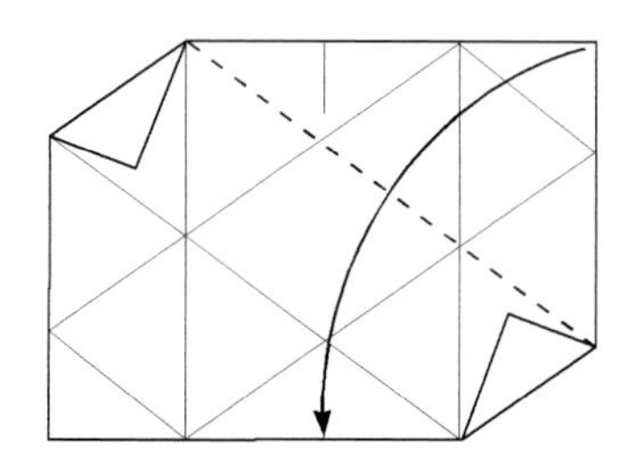

12

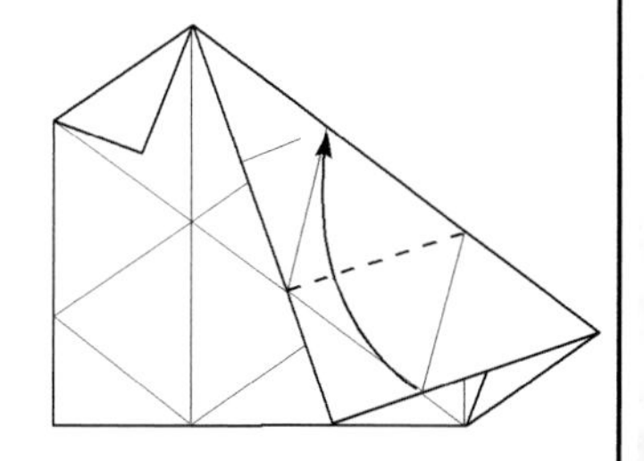

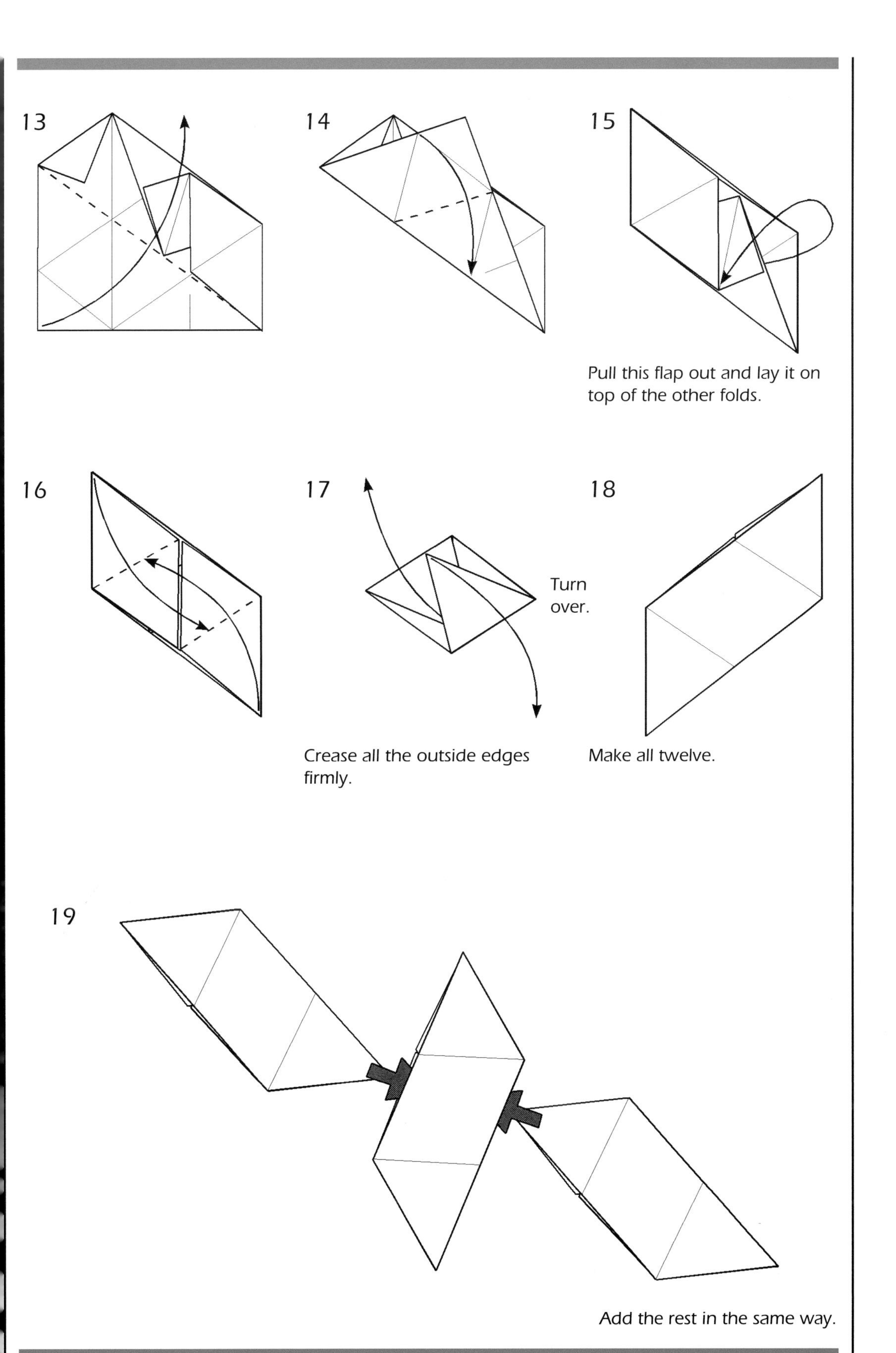
13
14
15
Pull this flap out and lay it on top of the other folds.
16
17
Turn over.
Crease all the outside edges firmly.
18
Make all twelve.
19
Add the rest in the same way.

Rhombic Pyramid

One sheet of A4 paper is required for each Pyramid.

Begin by folding the paper to step 9 of the Rhombic Dodecahedron, but flatten down the fold made in step 1 of those instructions to form a crease across the whole width of the paper.

1

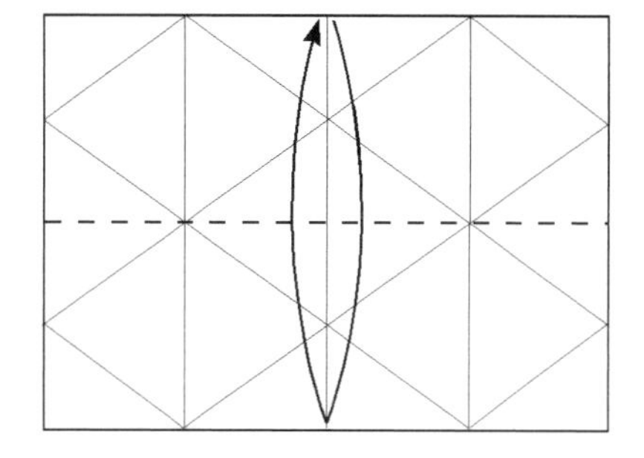

2

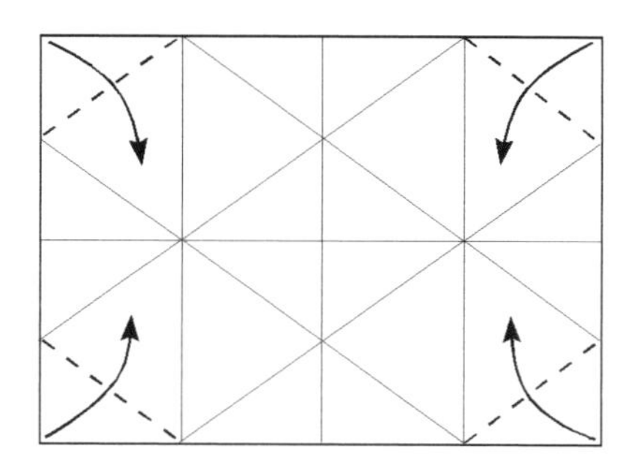

3

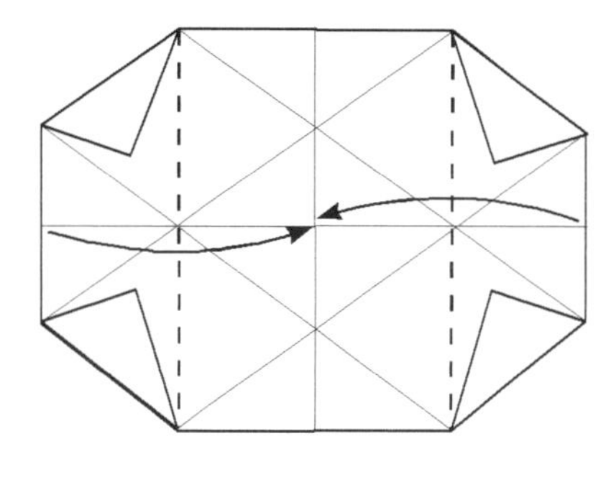

4

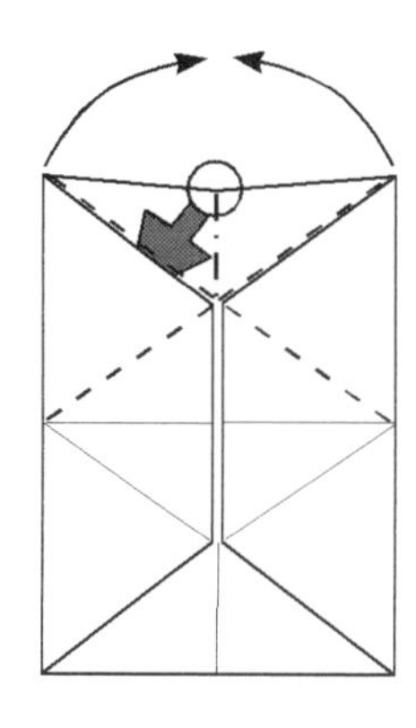

As you bring the two top corners together the model will become three dimensional and the circled point will form a flap.

This flap must be tucked sideways into the pocket, as shown, to lock the fold in place.

5

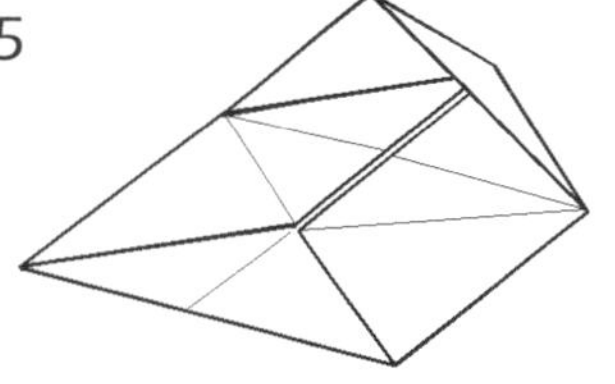

Crease firmly and undo back to step 4.

Turn the paper round.

6

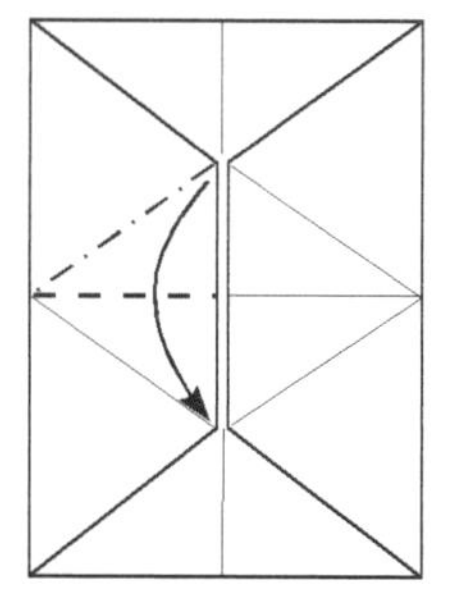

7

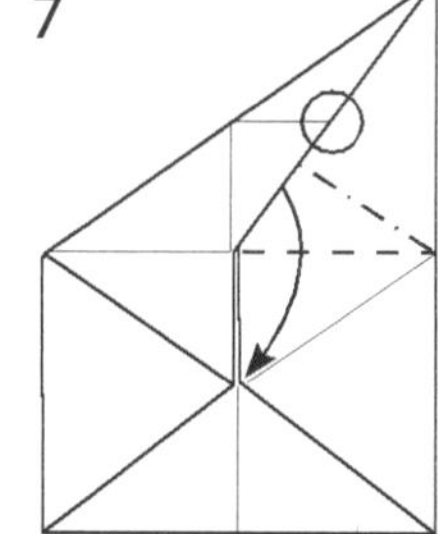

Repeat the same fold on the other half of the paper to form a loose pyramid.

The circled point will end up inside the model.

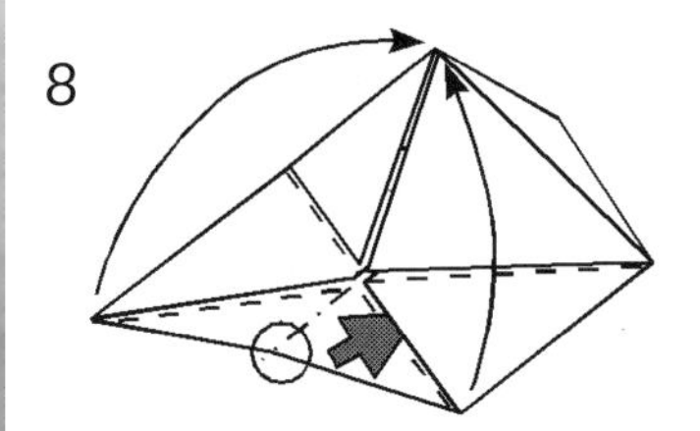

8

Refold the other end to lock the Pyramid in place.

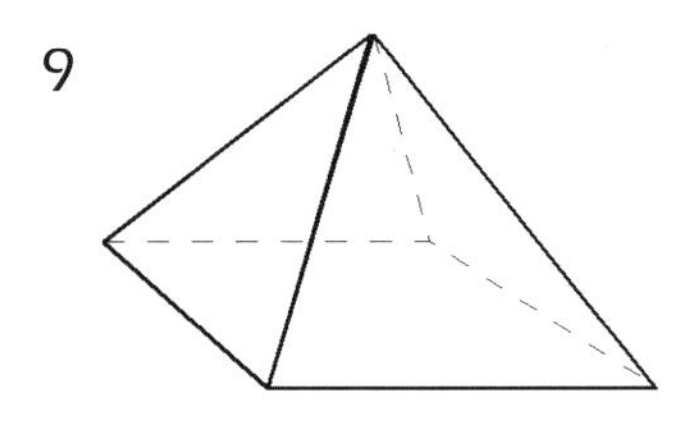

9

Complete.

Steps 6 to 8 are probably the most difficult folding sequence in this book. If you find them impossible you can still complete a similar - though less rigid - Rhombic Pyramid by simply repeating step 4 on the second half of the paper instead.

A compound Rhombic Tetrahedron can be made by first combining two Rhombic Pyramids into a Rhombic Octahedron and then surrounding it with four Rhombic Tetrahedra. To get the faces to scale, make the Tetrahedra from paper half the size of that used for the Pyramids.

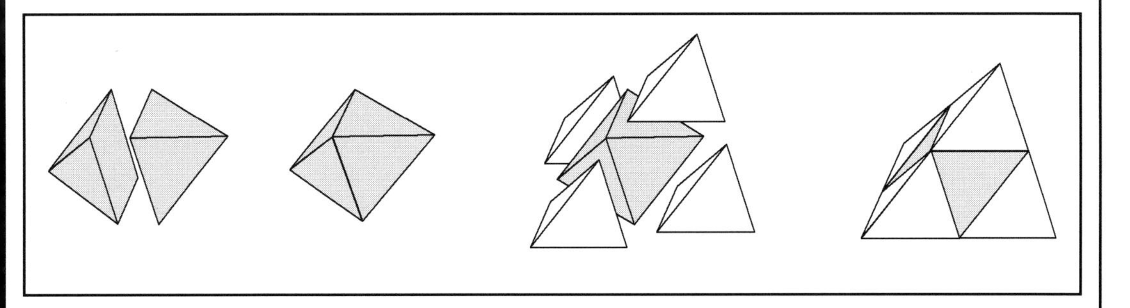

Four Rhombic Pyramids will also go together to make a Rhombic Tetrahedron.

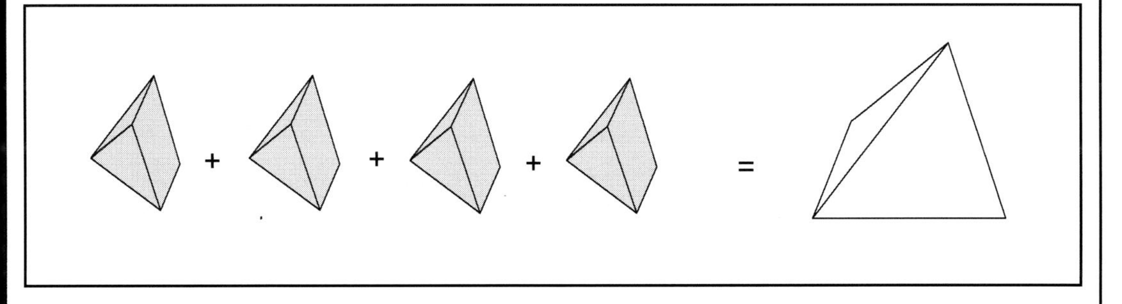

The solution is not immediately obvious when handling the Pyramids and makes an interesting manipulative puzzle.

Rhombic Tetrahedron

Two sheets of A4 paper are required.

This model is made in a very similar way to the regular Tetrahedron, but, because of the more acute angles, the assembly of the two modules is more difficult. It is probably well worth making the easier model first.

1

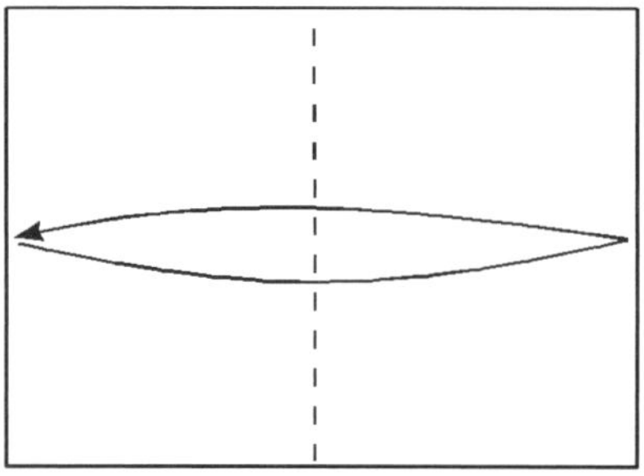

Turn over.

2

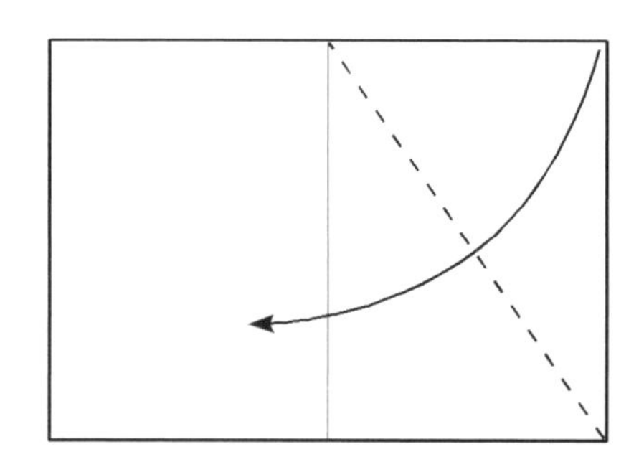

3

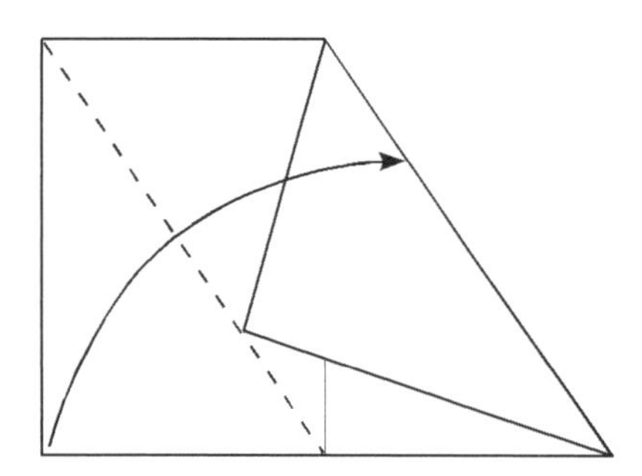

4

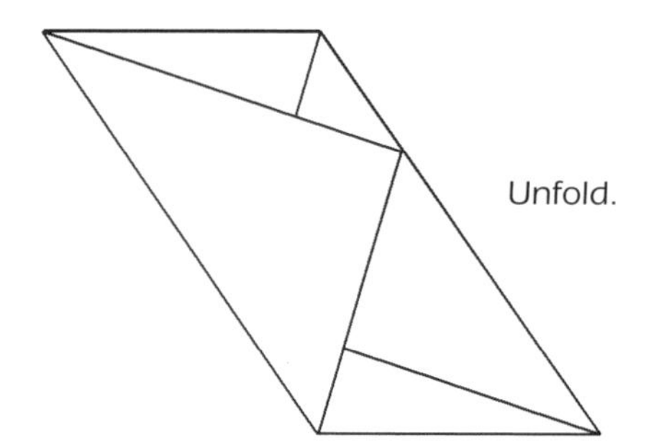

5

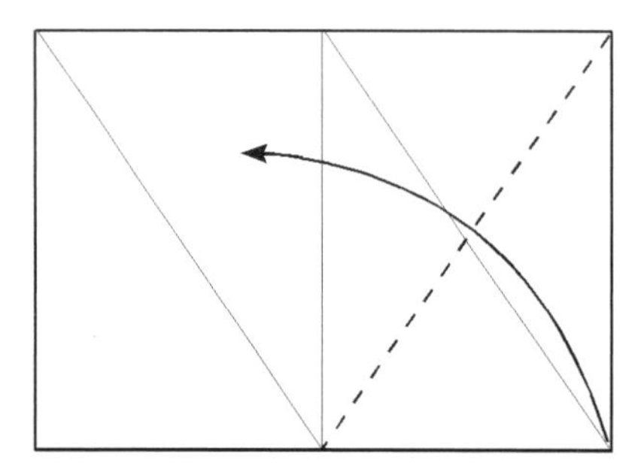

6

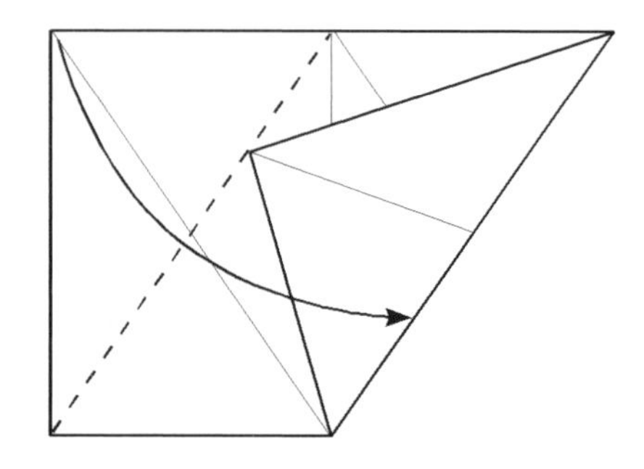

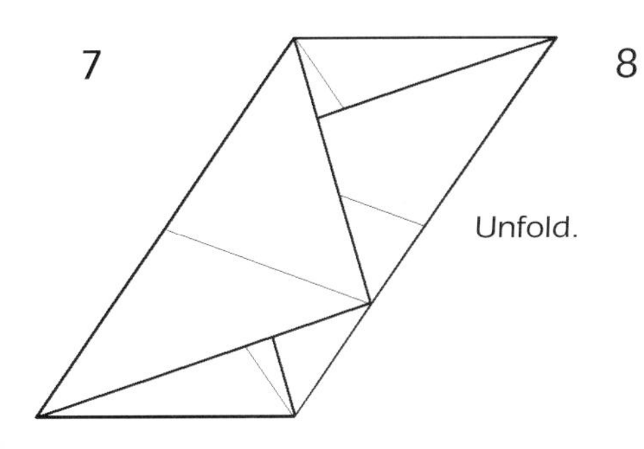

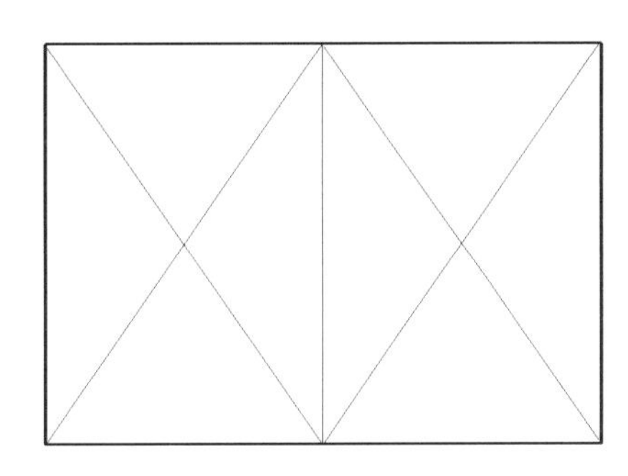

Fold both modules to this stage. From now on the second module is folded as a mirror image of the first.

Only the folding method for the basic module is shown.

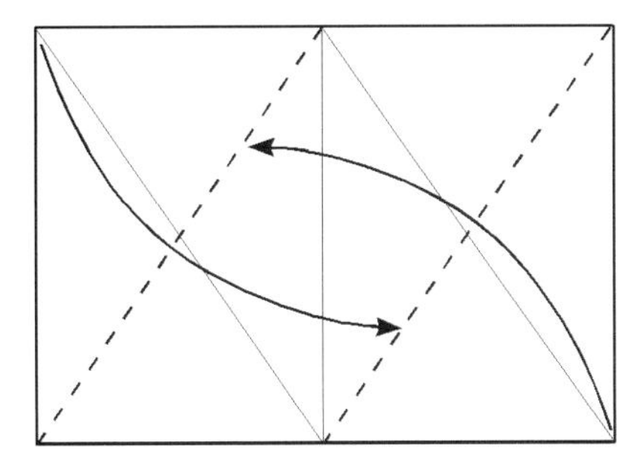

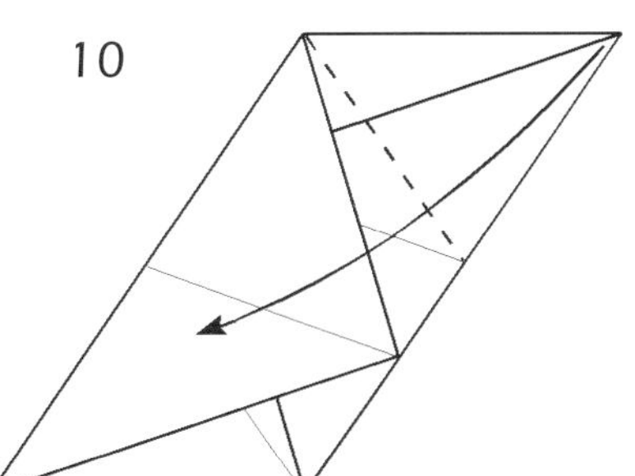

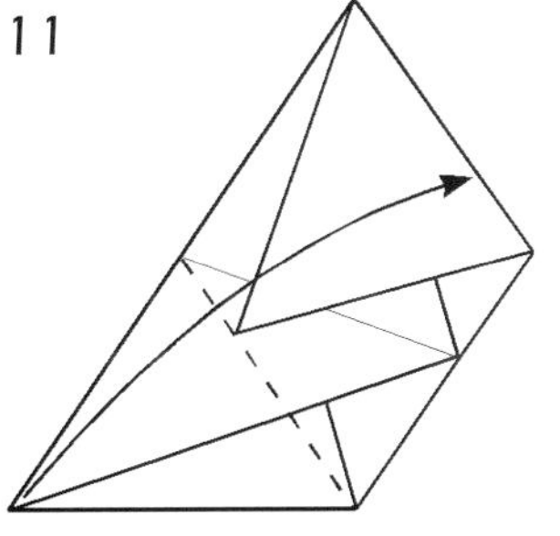

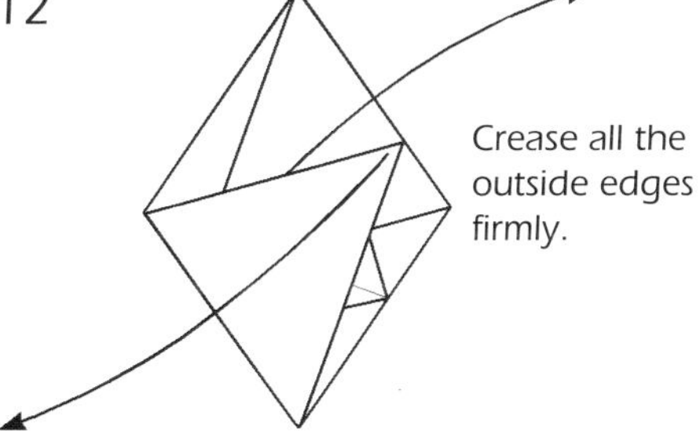

Crease all the outside edges firmly.

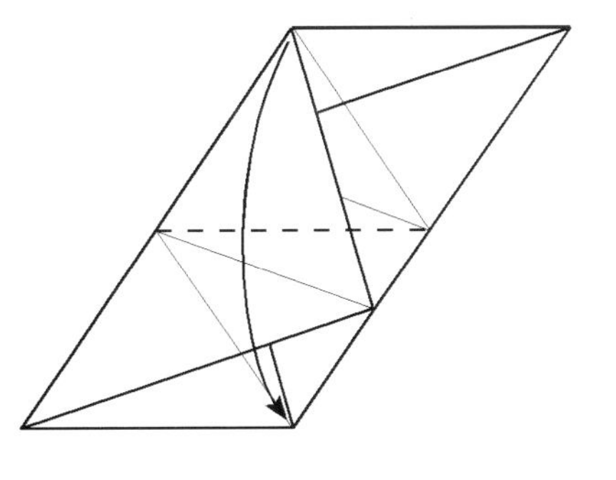

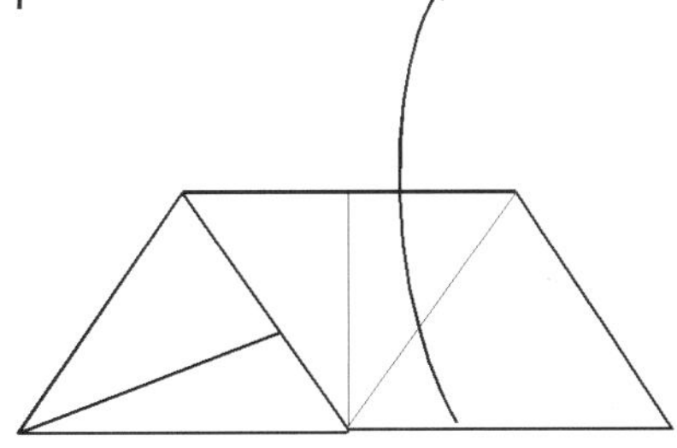

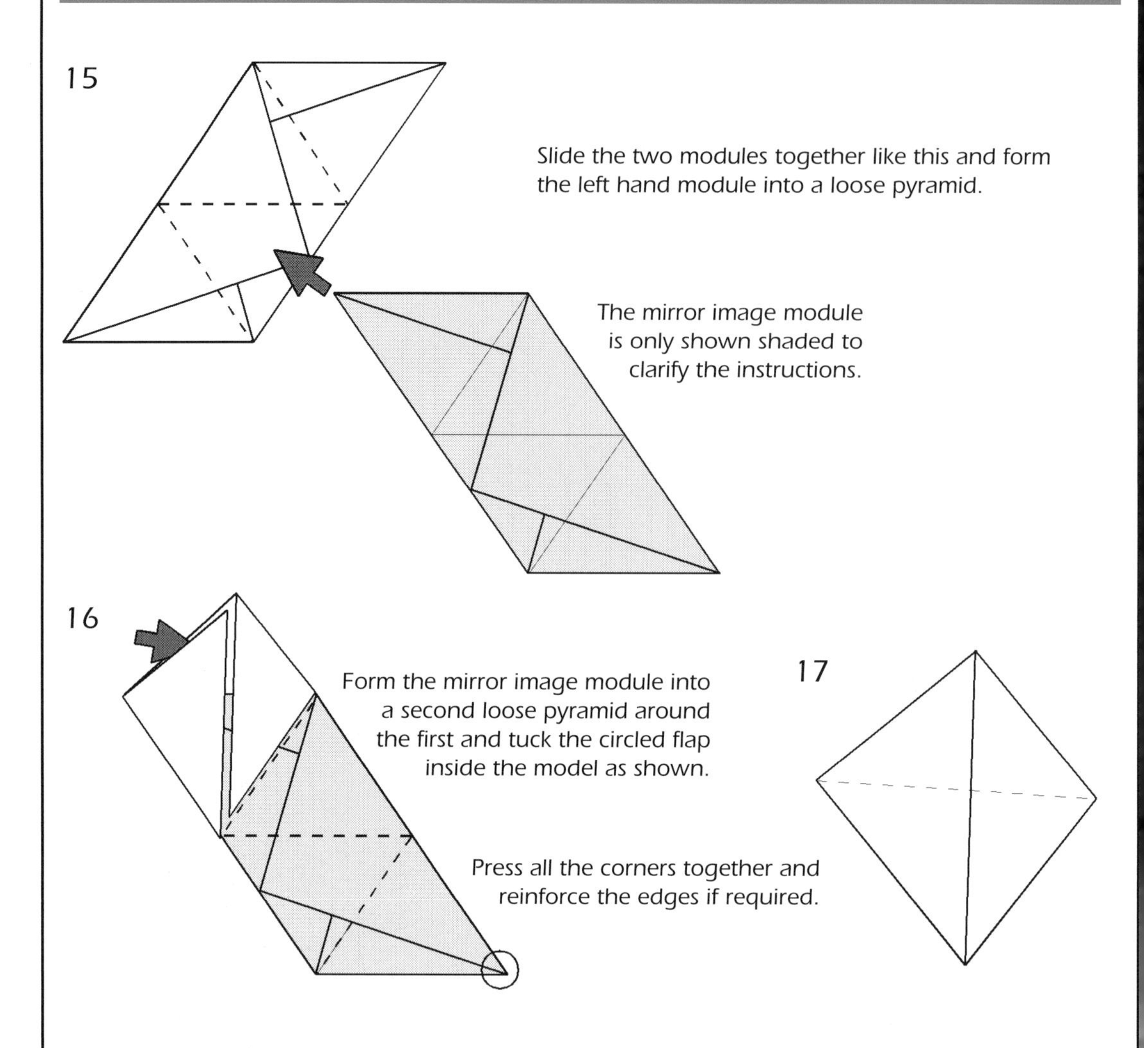

The Rhombic Tetrahedron is the common denominator of all the rhombic models in this collection, and has many interesting qualities. Like the Cube, it will pack space, and just as eight small cubes will pack together to form a large cube eight Rhombic Tetrahedra will pack together to form a large Rhombic Tetrahedron - although not in such a simple way.

The form is rhombic by analogy with the faces of the Rhombic Dodecahedron. Its four faces are formed of two identical rhombs folded across the shorter axis. Interestingly, if you cut the Rhombic Tetrahedron into two equal halves through the centres of four of its edges - see diagram - the plane formed by the cut is also an identical rhomb. If you cut through the regular Tetrahedron in the same way the resulting plane is, of course, a square.

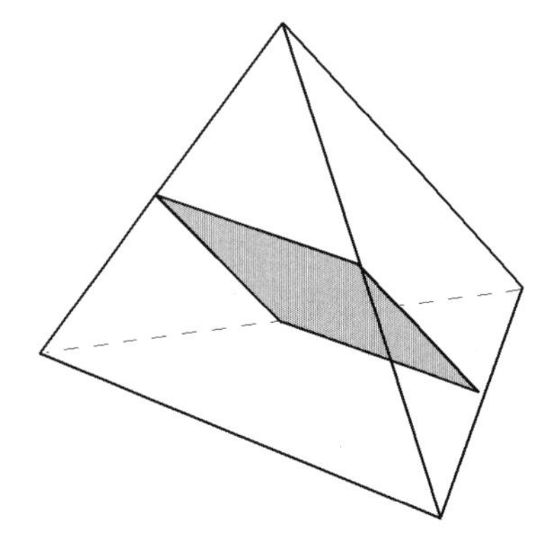

Ring of Rotating Rhombic Tetrahedra
Decorative Rhombic Dodecahedron
Rhombic Star

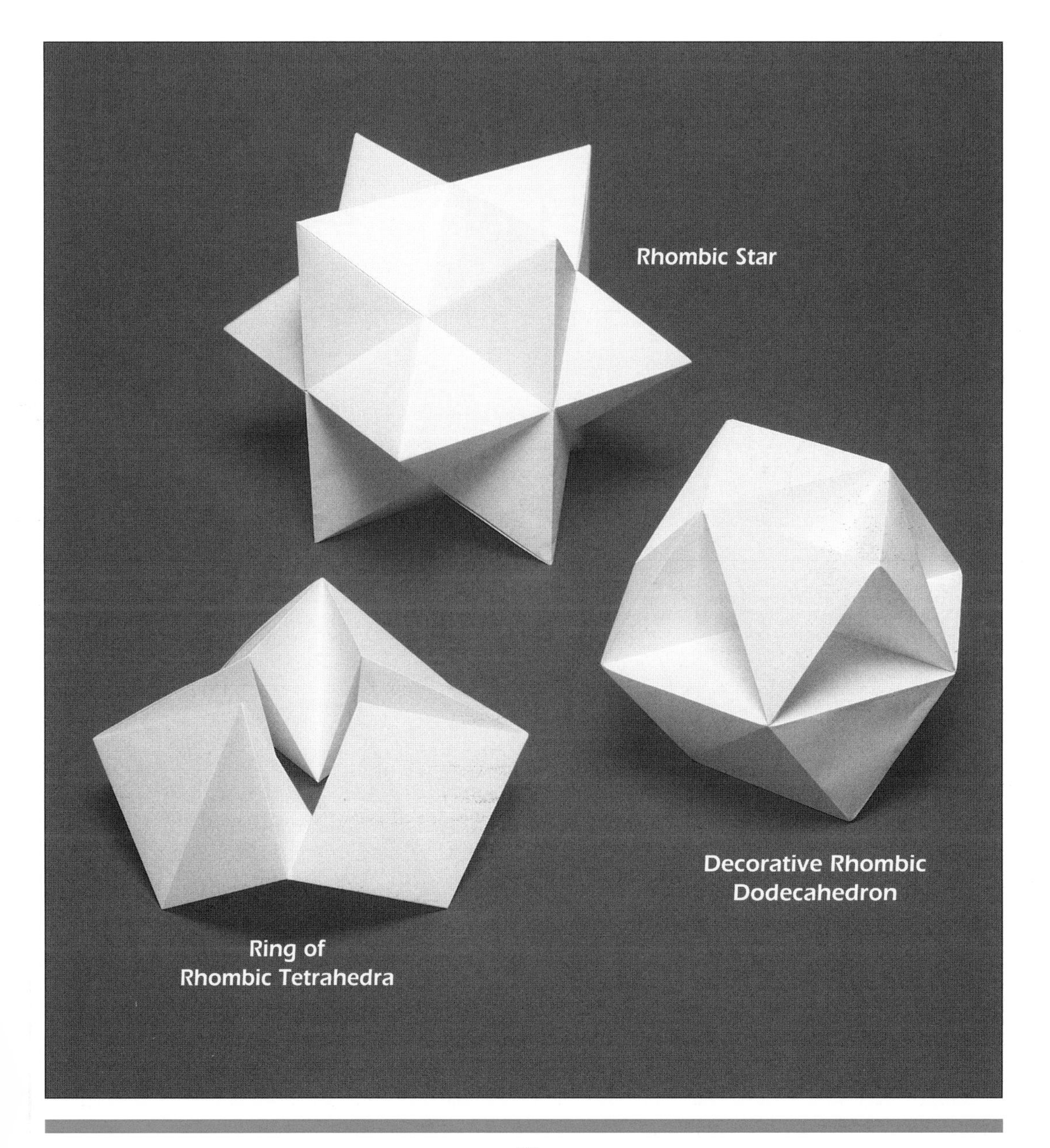

Ring of Rotating Rhombic Tetrahedra

Despite its accurate but rather uninspiring name this model is simple to fold and assemble, and the result is an appealing toy.

Four sheets of A4 paper are required.

1

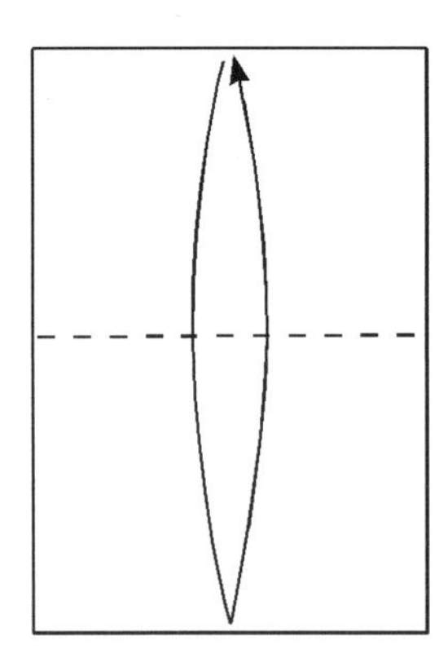

2

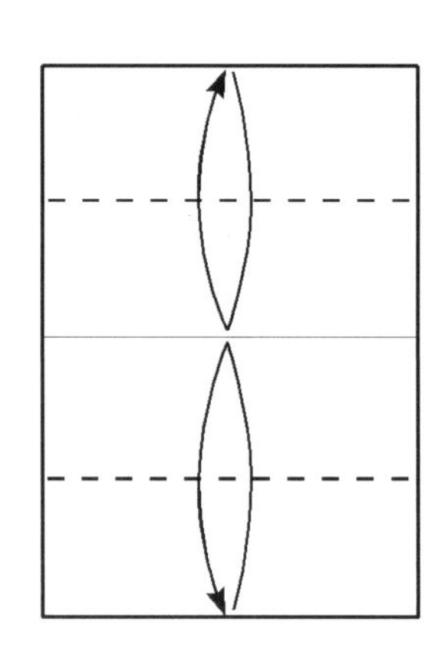

3

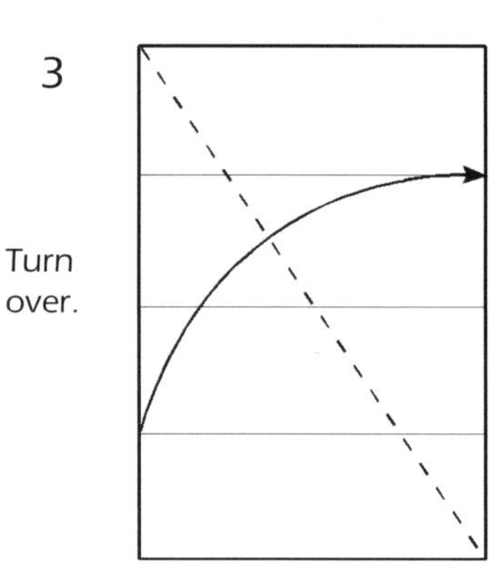

4

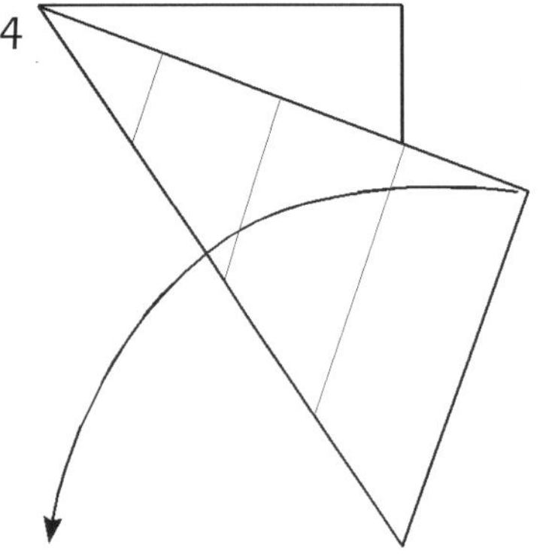

5

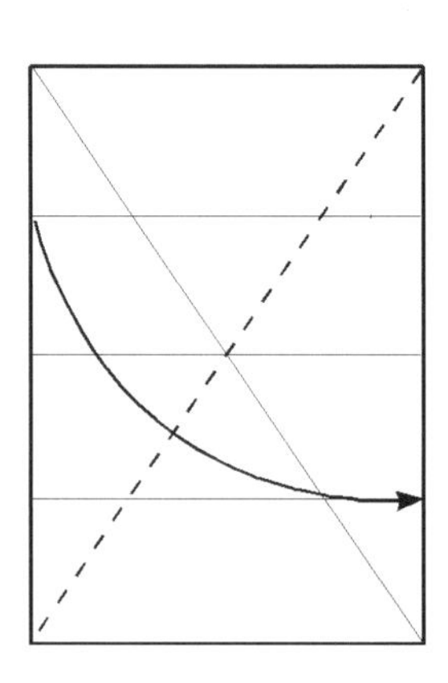

6

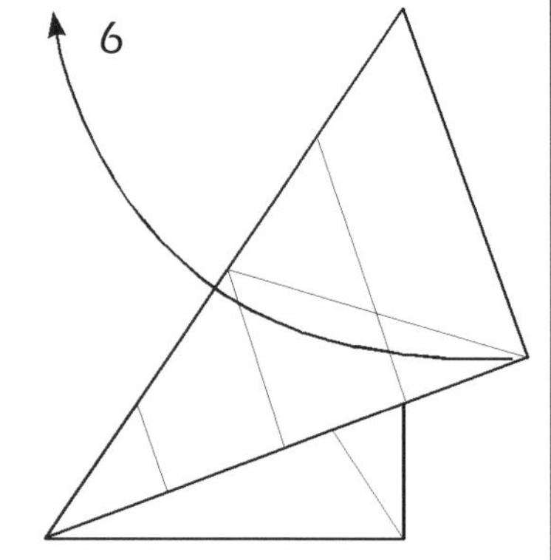

7

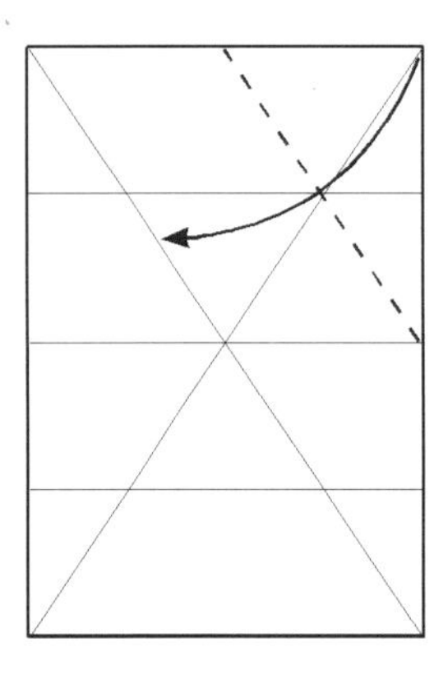

8

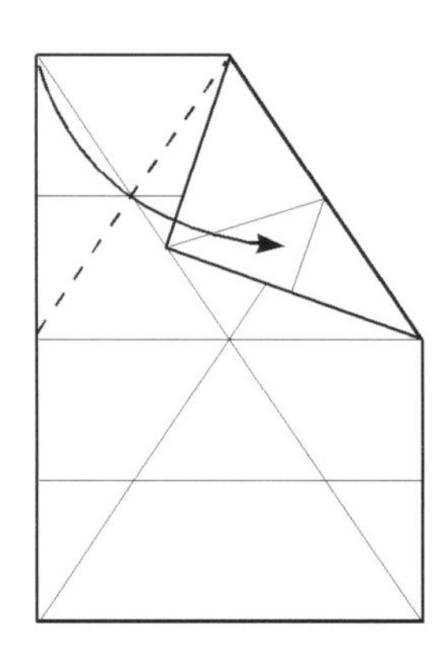

9

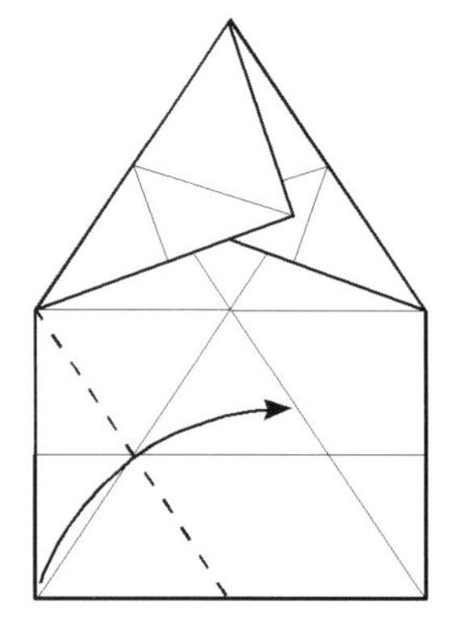

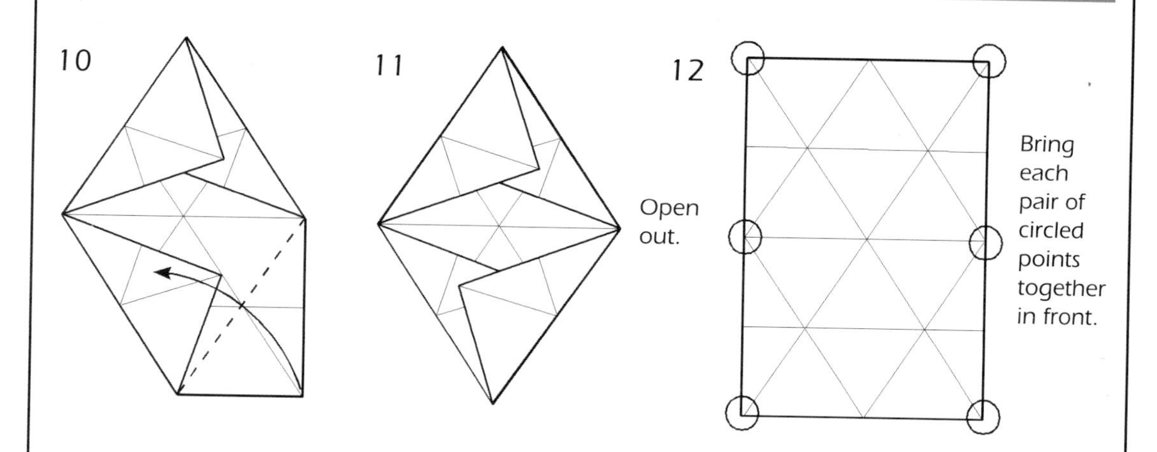

10

11

12

Open out.

Bring each pair of circled points together in front.

13 As long as you remembered to turn the paper over between steps 2 and 3 the model will collapse into a three-dimensional form, like this:

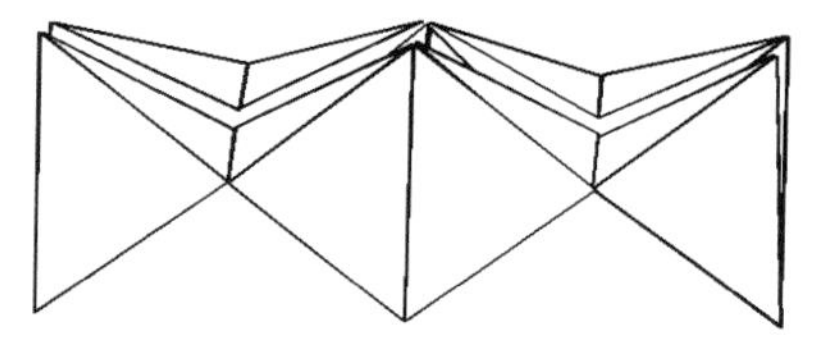

Make all four.

14 Put the four modules together like this.

The shaded halves go inside the white halves of the other modules.

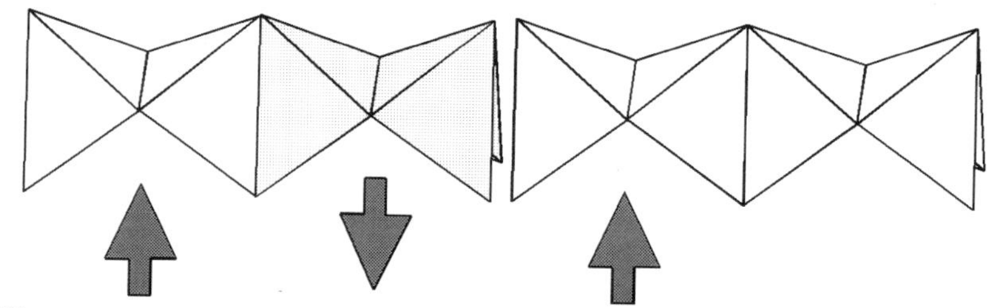

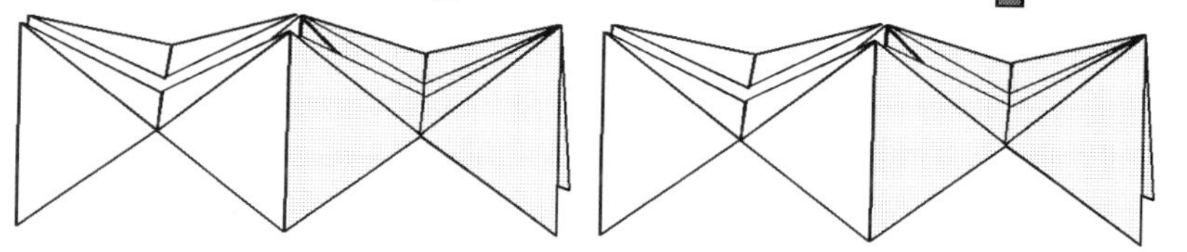

15 This caterpillar-like form is the result.

Curl the model up and insert the shaded part of the module on the right inside the open end of the module on the left.

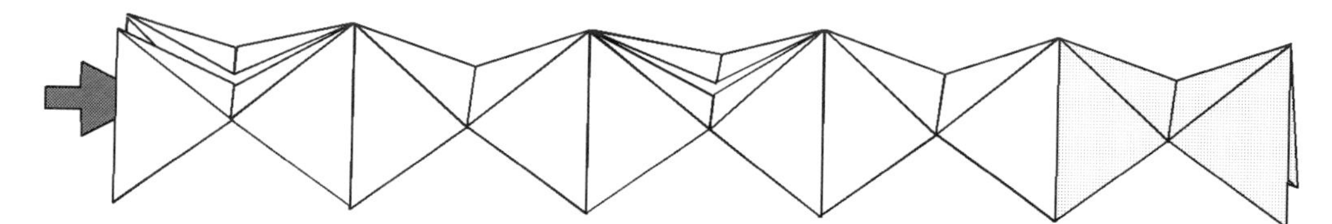

Once this is done the rotating ring is complete.

The Tetrahedra rotate through the centre of the ring. After a few rotations the folds will settle down and the ring will turn without effort.

Decorative Rhombic Dodecahedron

This model is a fine example of an elegant but unusual form that occurs naturally when you begin to play with the possibilities inherent in modular origami. There are many ways of making it, but this is the simplest.

Six sheets of A4 paper are required.

The six modules form a cladding around a Skeletal Cube made from quarter A4 paper.

Begin by folding each sheet to step 11 of the Ring of Rotating Rhombic Tetrahedra.

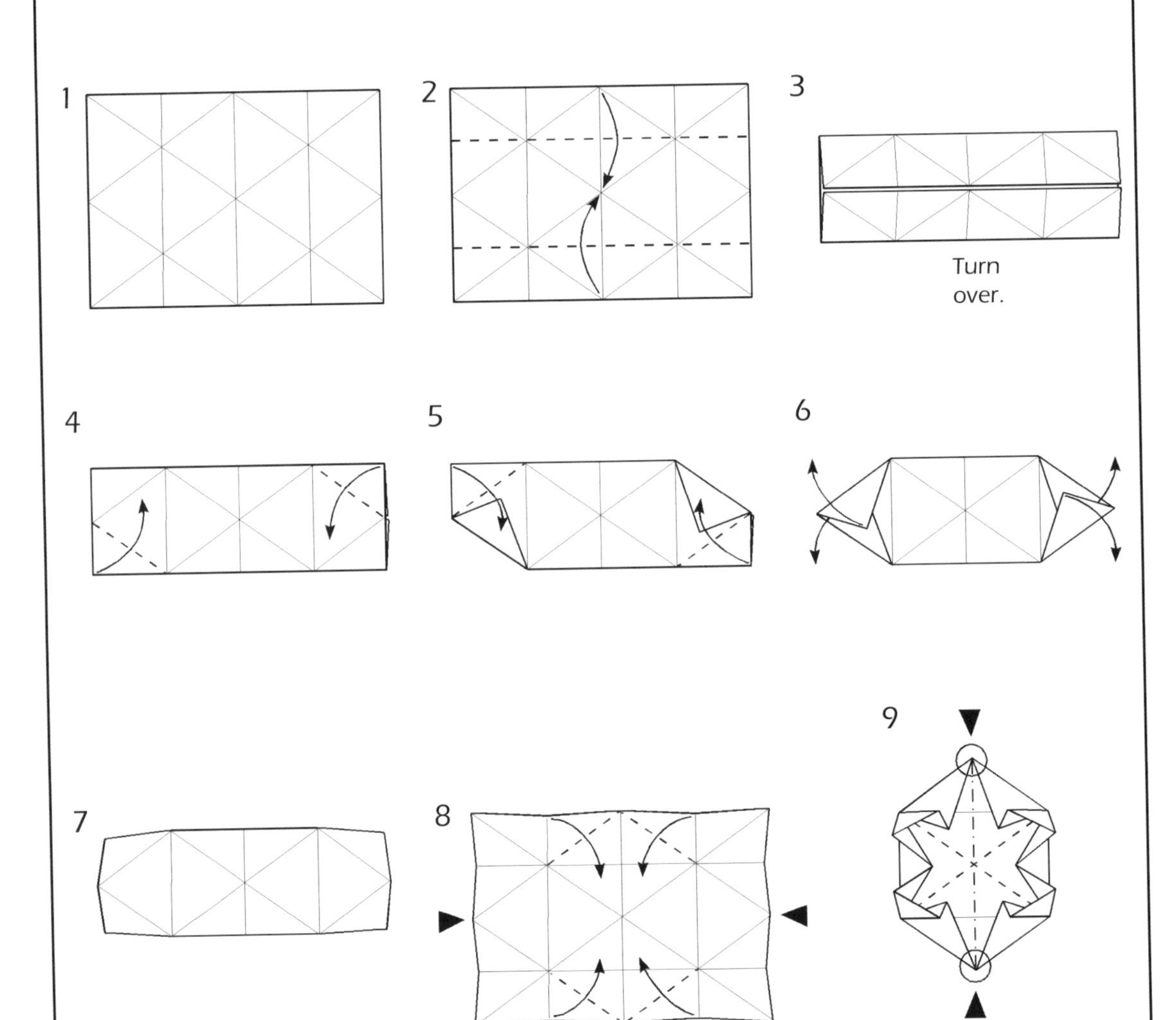

Turn over and open out completely.

Allow the paper to remain slightly three-dimensional.

Bring the two circled points together so that the centre sinks.

Turn over.

10

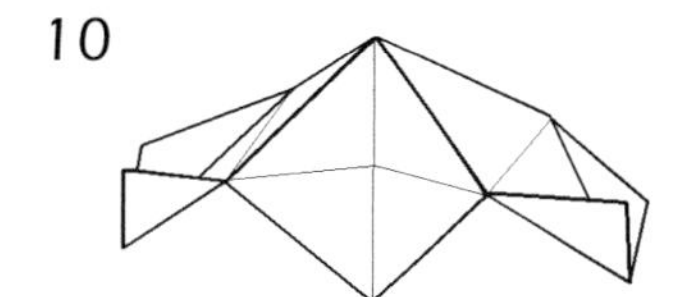

The finished module.
Make all six.

11

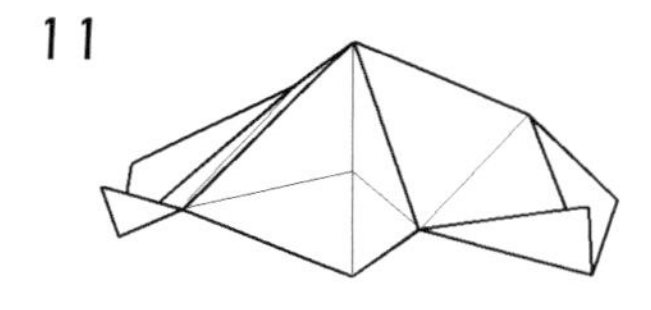

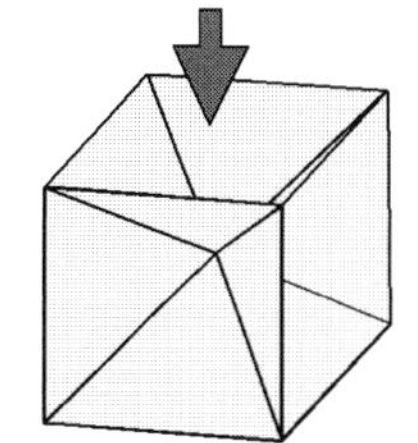

12

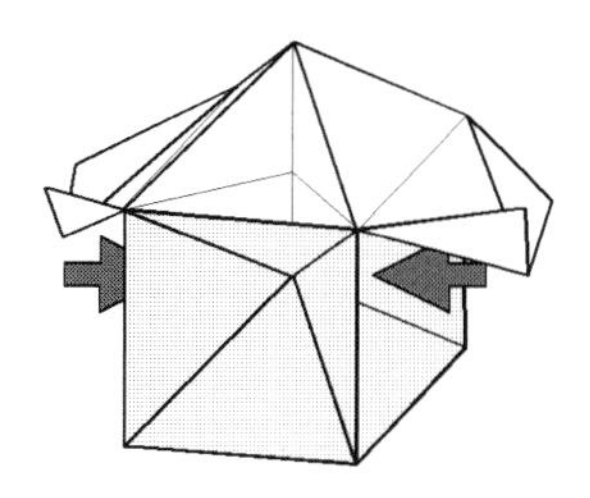

Tuck the flaps away inside the model.

13

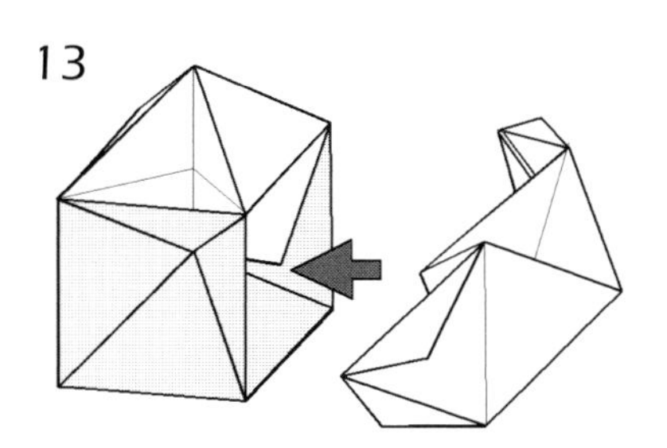

The body of the second module holds the flaps of the first in place.

14

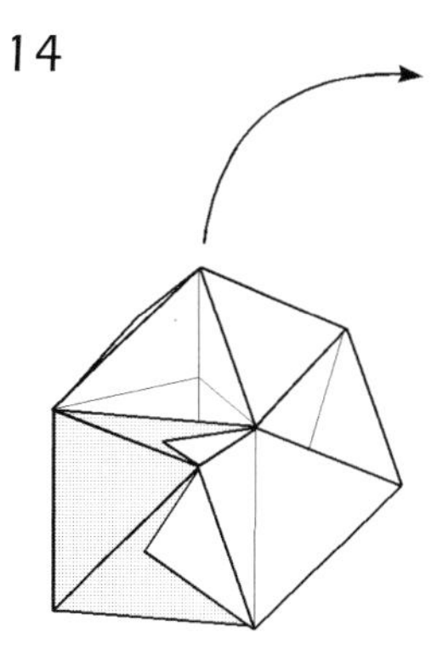

Lift up this module to add the third and fourth modules at front and back, then lower it down again to lock their upper flaps in place.

Add the fifth module.

You will discover that it is impossible to add the last module in this way because the module you need to lift up to insert the final flap is held securely in place by the others.

Fortunately though, you can still complete the model by curling the flap around and slipping it gently down into the gap between the underlying skeletal cube and the cladding. This is much easier if you start sliding the flap in as near to the corners of the underlying cube as you can.

Rhombic Star

Mathematically, and also in the way it is constructed, this attractive model is a Stellated Rhombic Dodecahedron.

Begin by making modules for the Rhombic Dodecahedron explained earlier in this book. Twelve more sheets of A4 paper are required for the additional units.

1

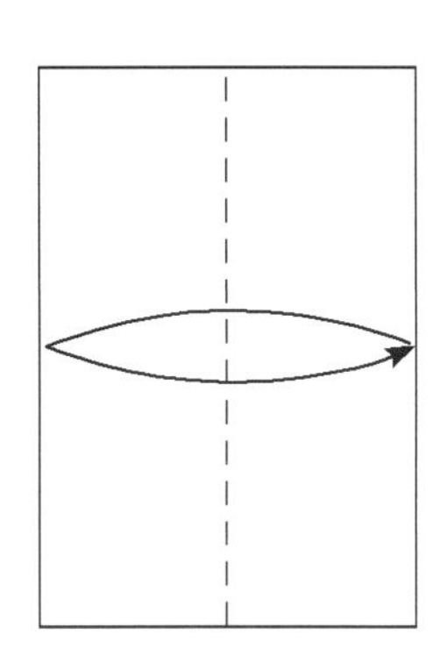

2

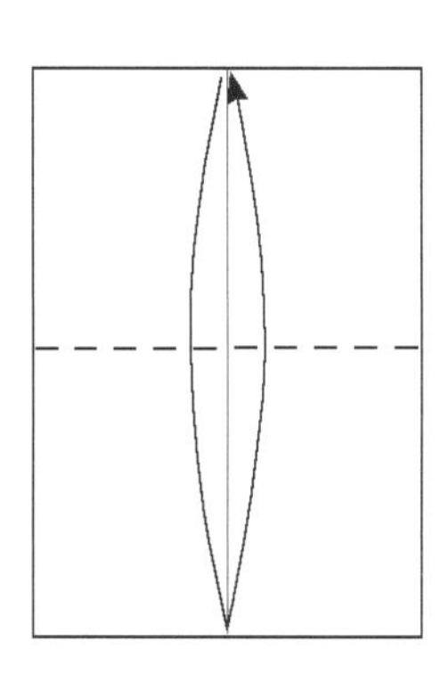

3

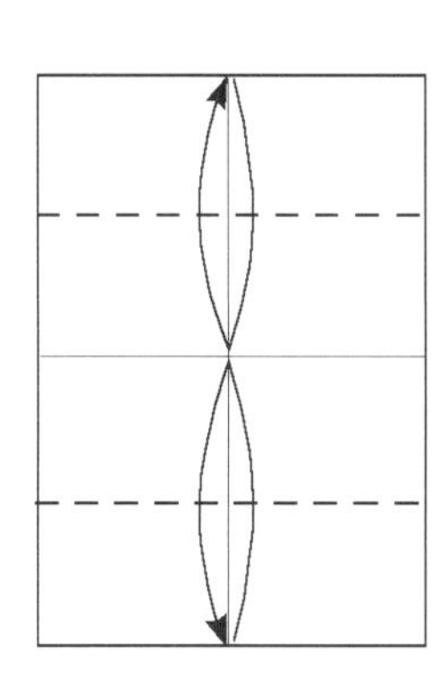

4

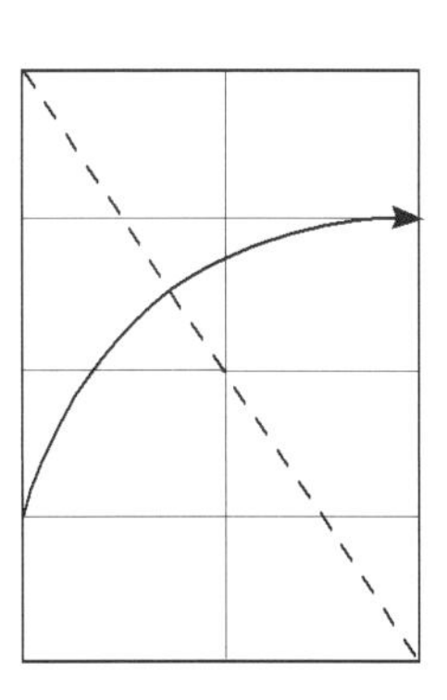

5

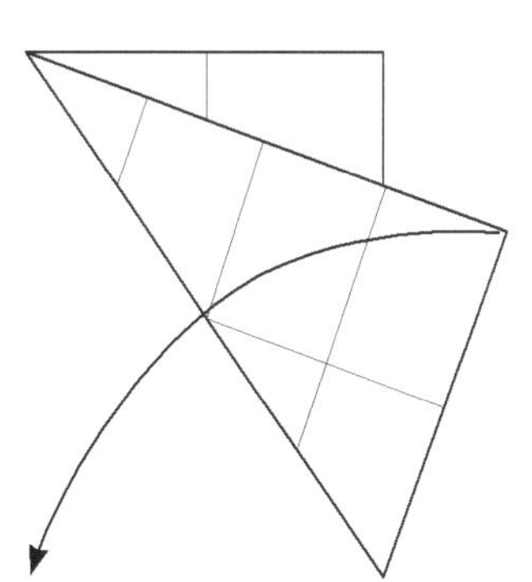

6

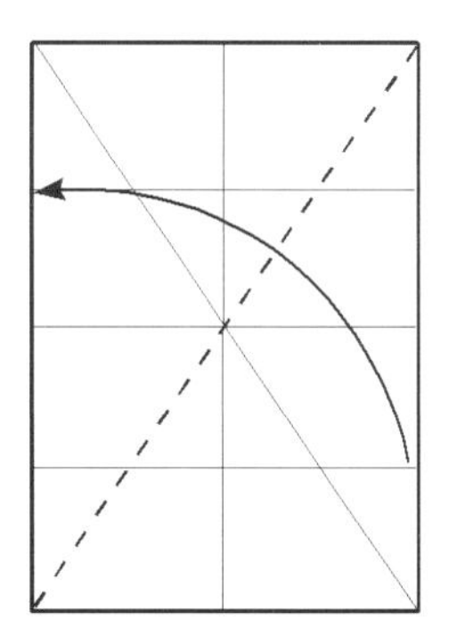

7

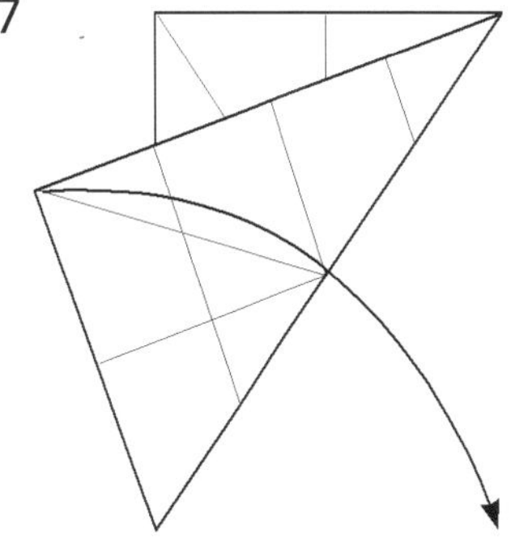

8

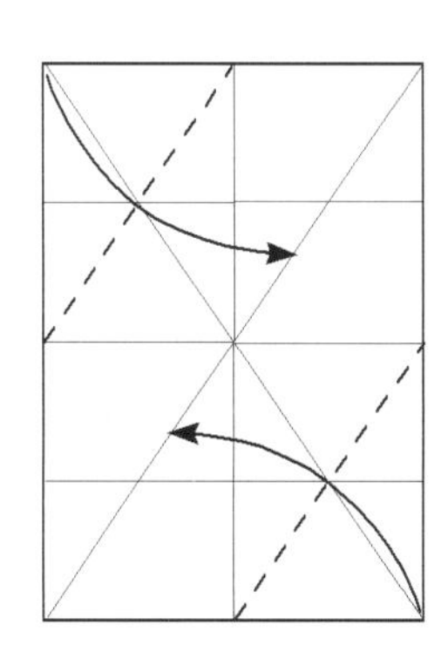

9

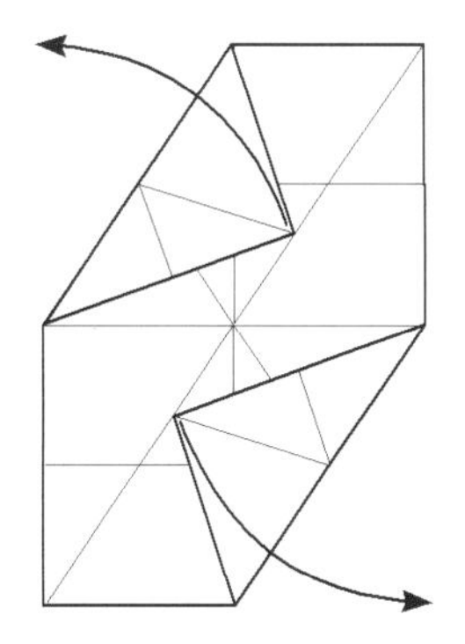

10

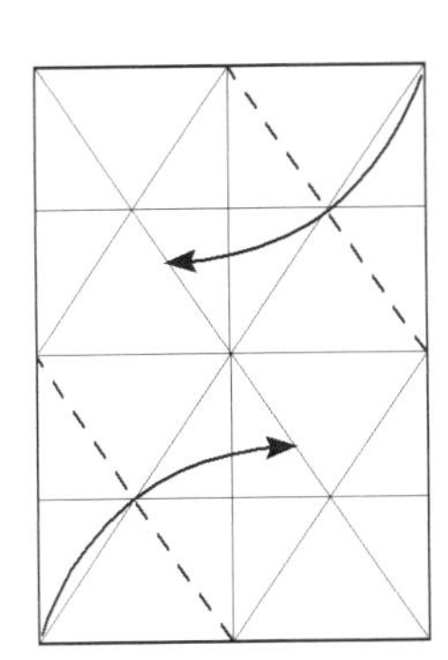

11

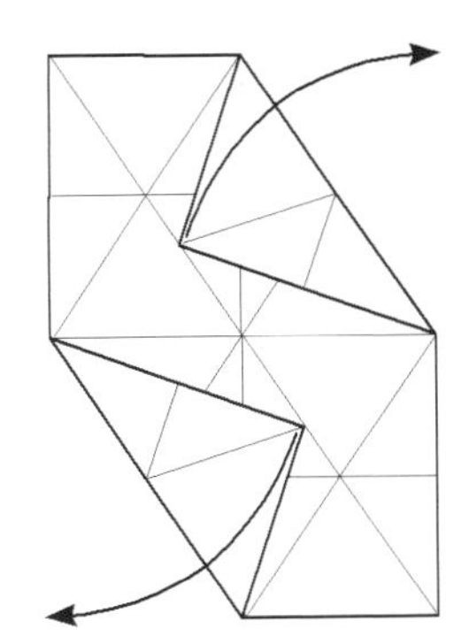

12

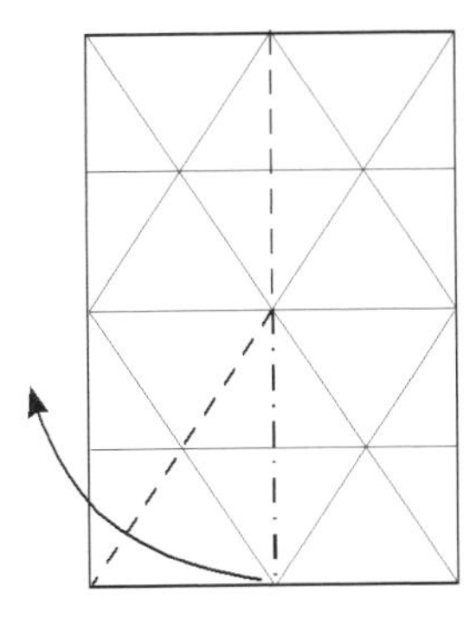

As you make this fold the model will become three-dimensional.

13

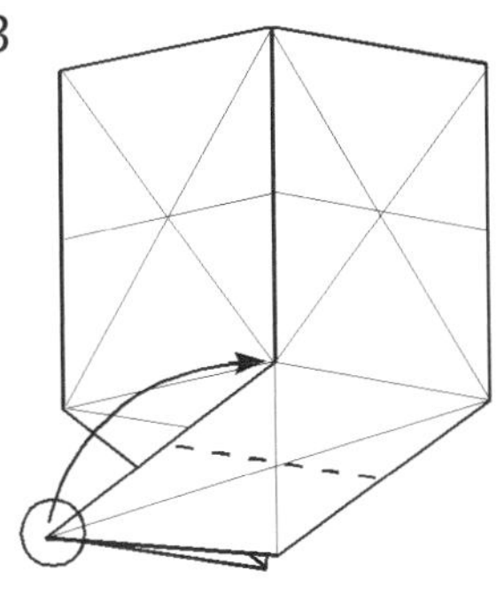

Make this fold in the top layer of the paper only, ensuring that the circled corner goes right into the centre of the model. As you do this the small flap hidden underneath will swivel up and across to the left, and can be flattened down into the position shown in diagram 14.

14

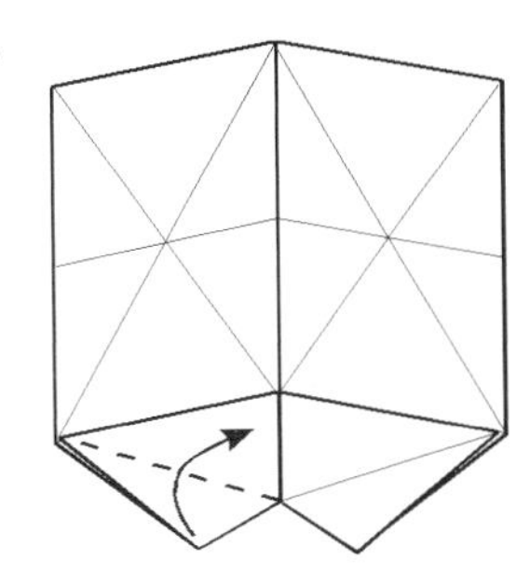

This is the result.

15

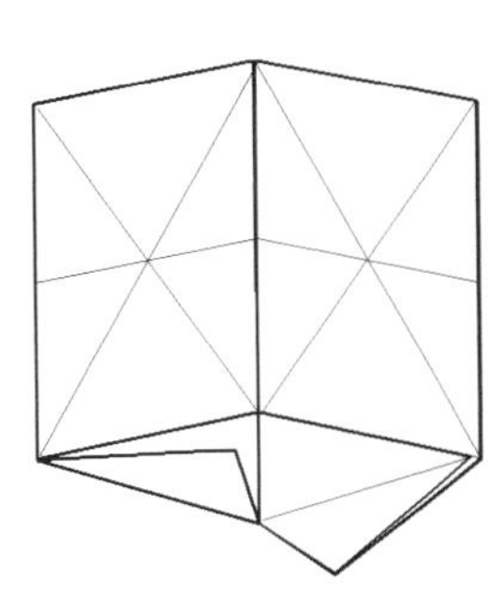

Undo the last few folds, turn the paper around and repeat folds 12 to 14 on the other end of the paper.

Then refold the first end as well.

16

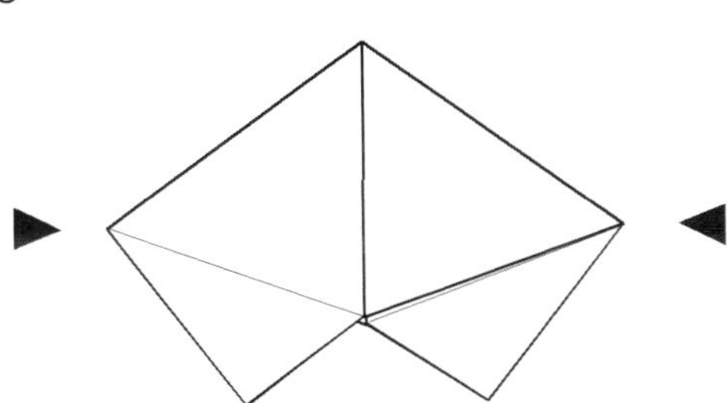

Crease all the outside edges firmly then apply gentle pressure to both sides. The module will open up into a pyramid, then squash flat in the opposite direction.

17

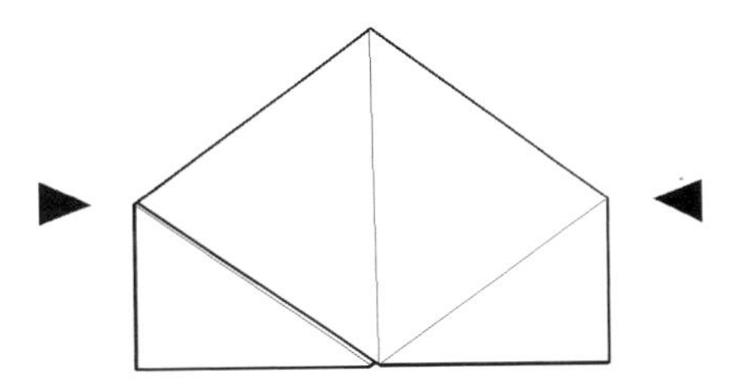

Crease the two top edges firmly, then open it up so that it becomes a rhombic pyramid again.

18

Each rhombic pyramid fits onto one of the modules from the Rhombic Dodecahedron like this.

The two flaps concealed inside the pyramid go into the sockets on the base module as shown.

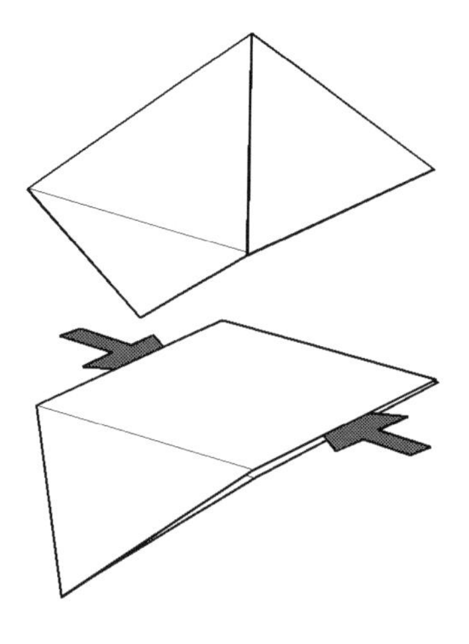

19

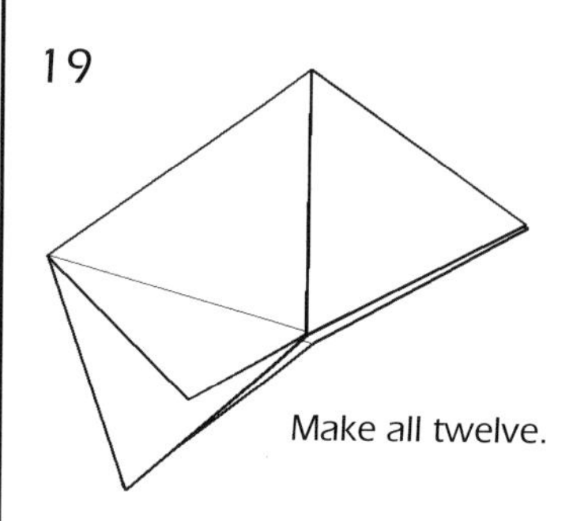

Make all twelve.

From here on the model is assembled in exactly the same way as the Rhombic Dodecahedron.

Alternatively it is possible to add the pyramids to the base modules after they have been assembled into the Rhombic Dodecahedron. In this case, if you begin by adding just four pyramids in any straight line around the Dodecahedron the model first turns into a Rhombic Octahedron.

Dodecahedron
Outline Dodecahedron

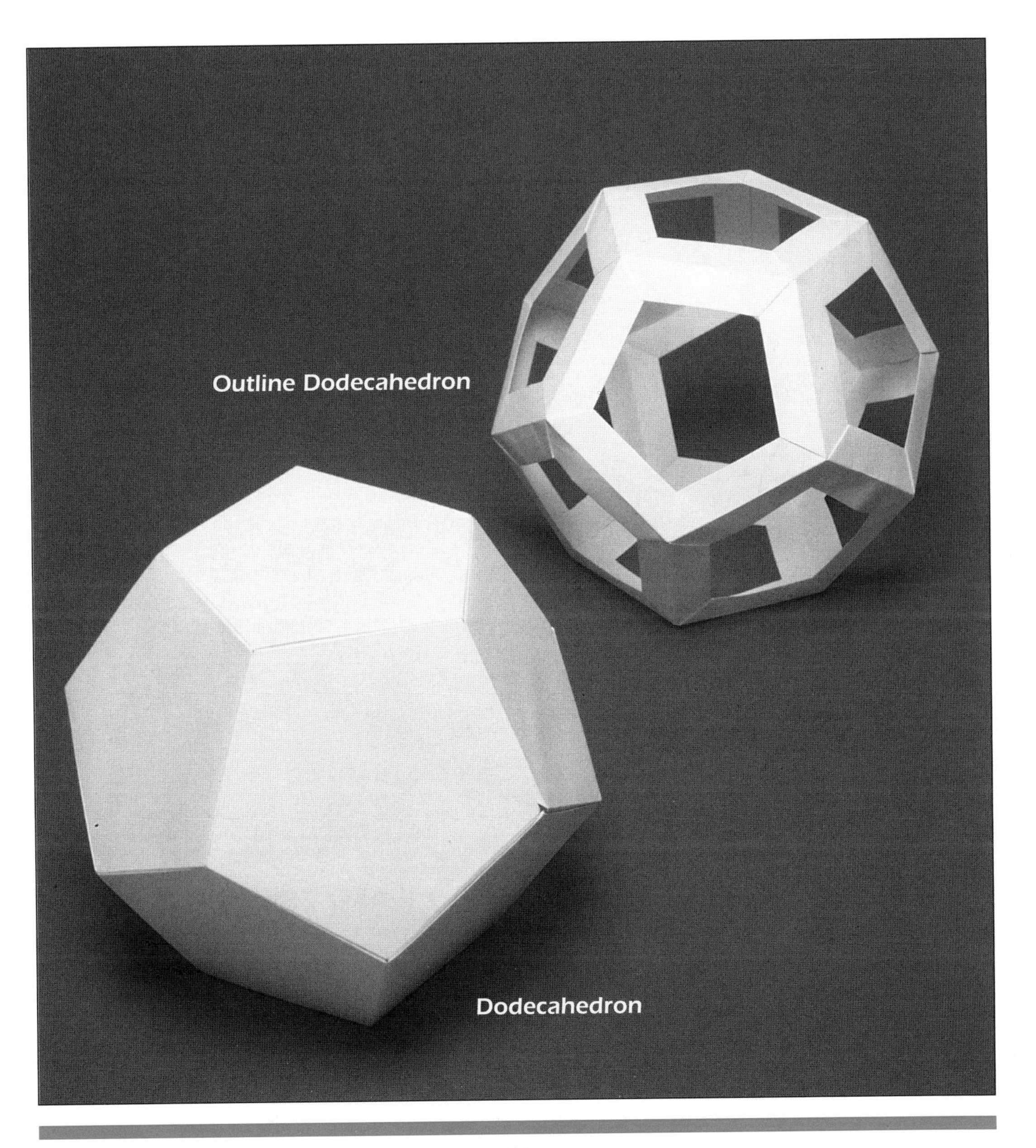

Dodecahedron

Twelve sheets of A4 paper are required.

This is by far the most difficult model in this book and complete accuracy is essential if a good result is to be obtained.

Begin by preparing each sheet to step 5 of the Cube.

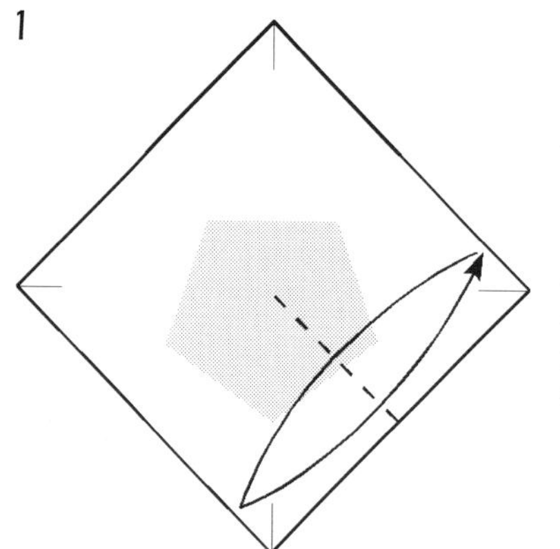

Try not to crease the paper inside the central shaded area.

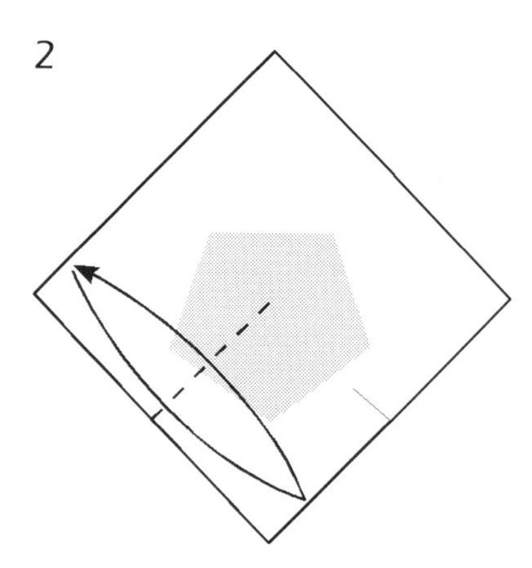

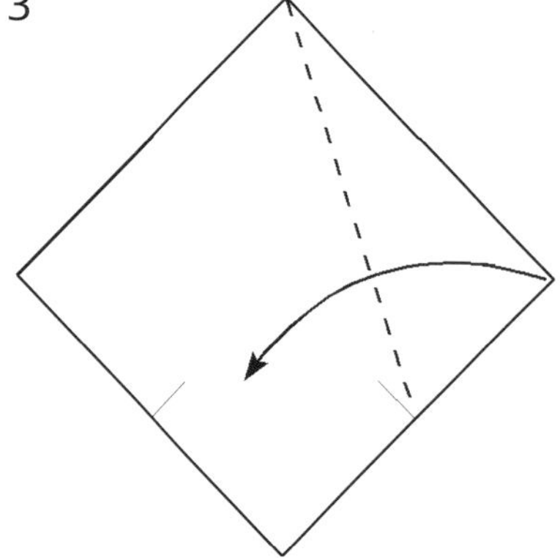

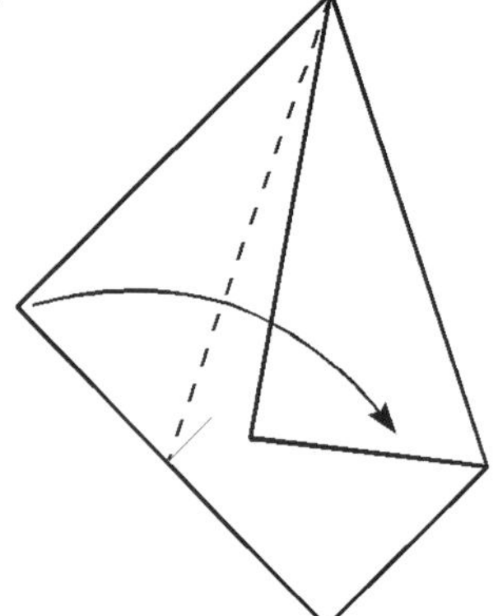

5

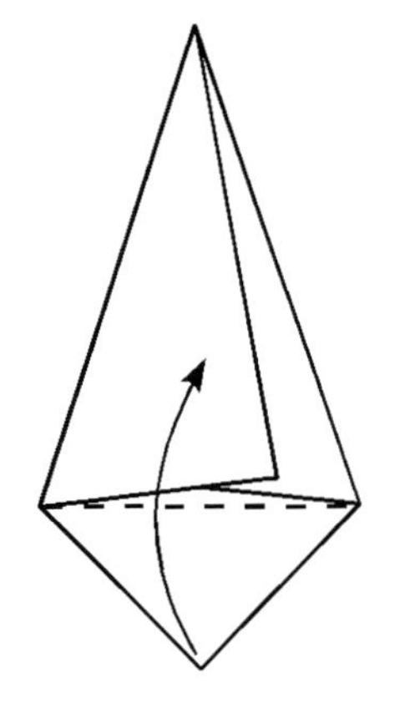

6

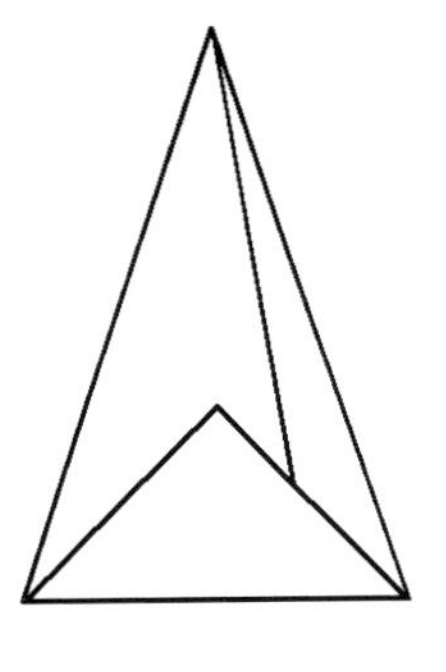

Unfold.

7

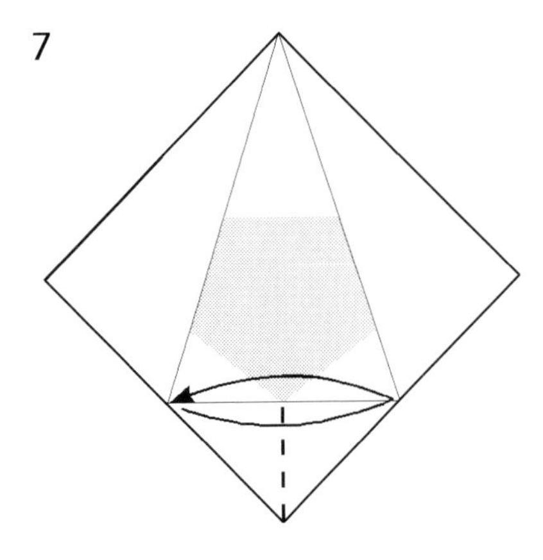

8

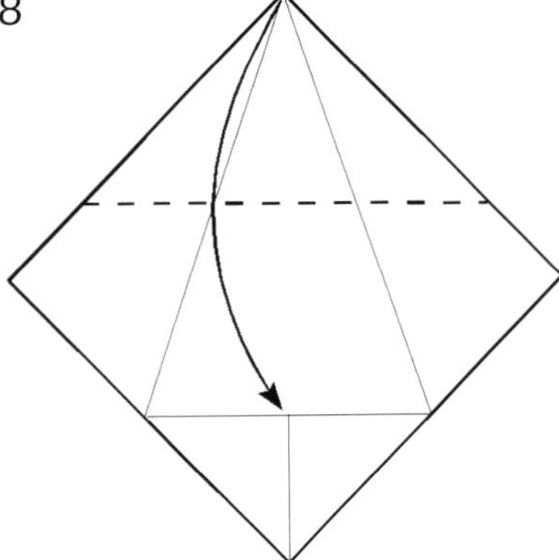

9

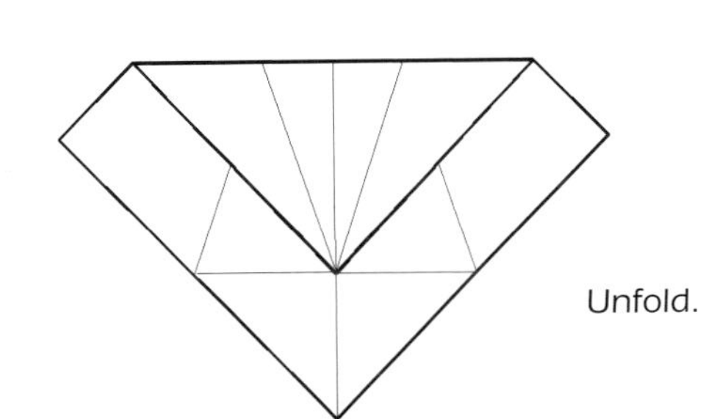

Unfold.

10

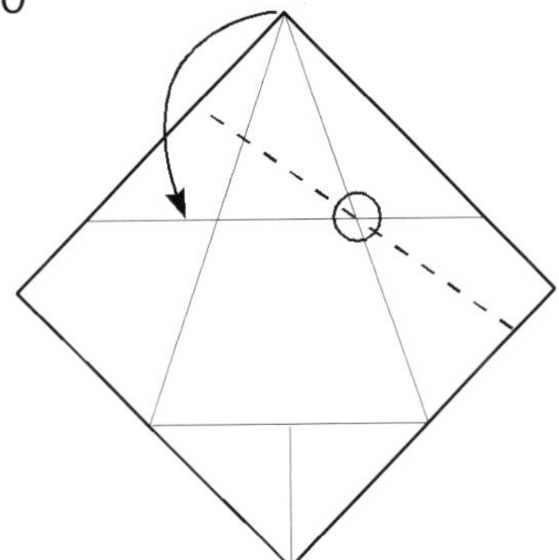

This new fold must pass exactly through the point where the two creases cross.

11

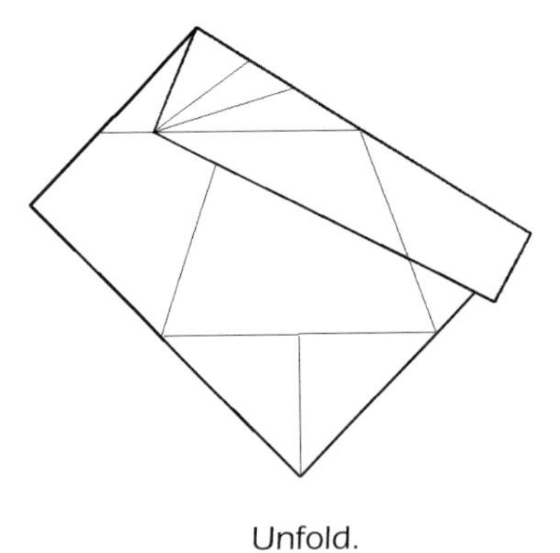

Unfold.

12

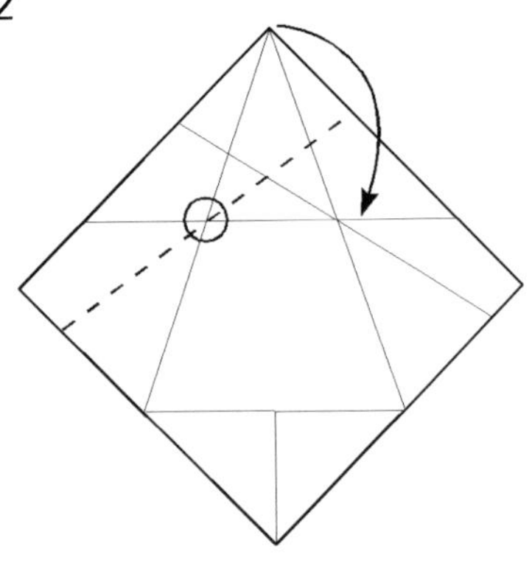

This new fold must pass exactly through the point where the two creases cross.

13

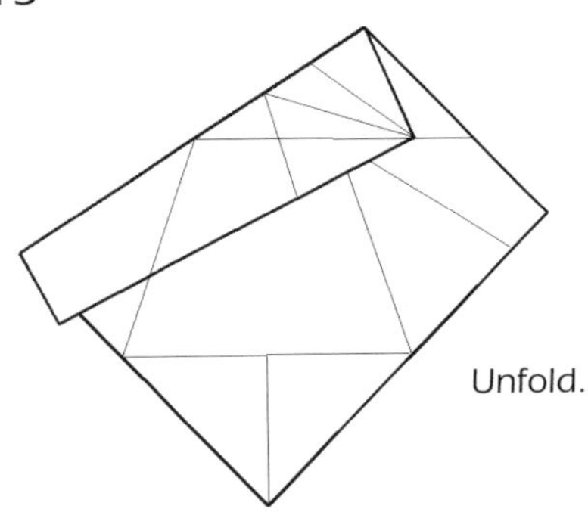

Unfold.

14

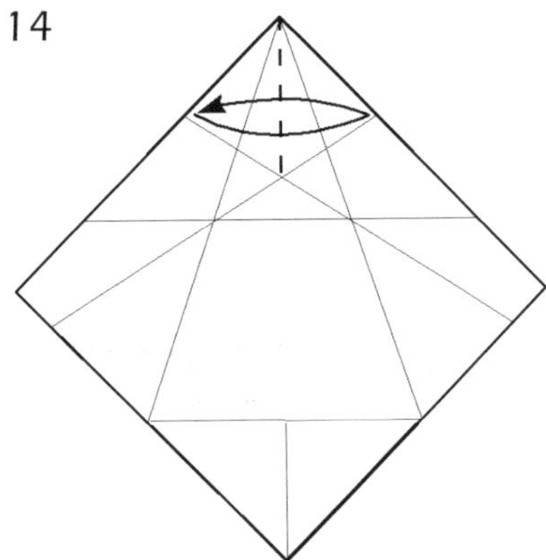

15

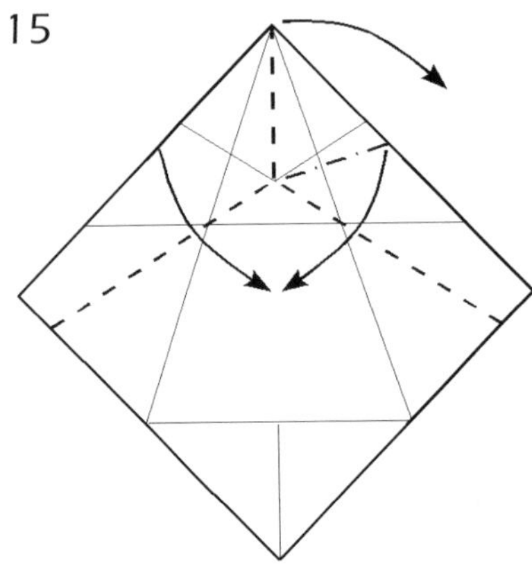

Fold both sides in together and squash the top sideways. A new crease will form in the position shown.

16

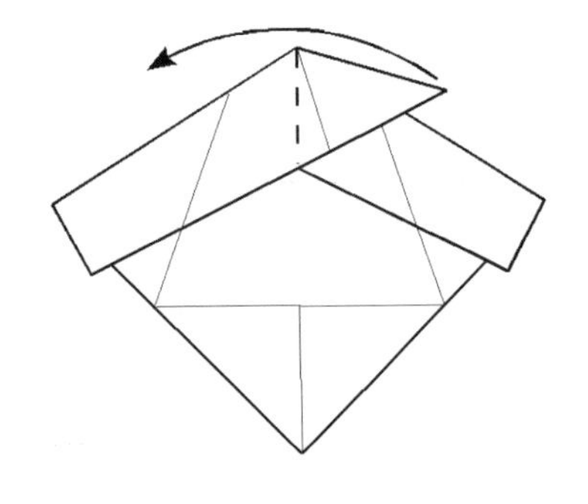

17

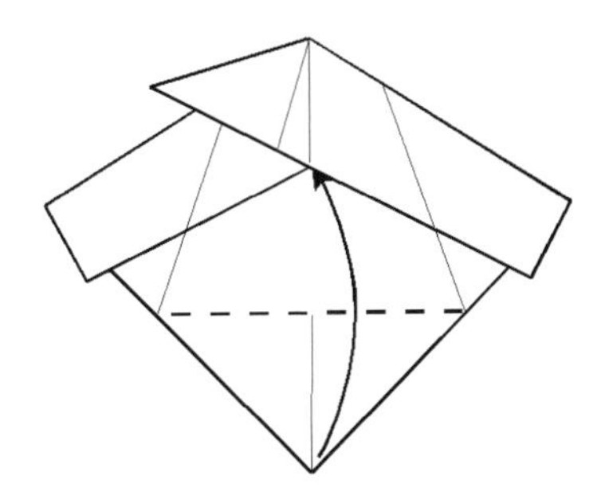

Tuck the point of the flap underneath the upper folds.

18

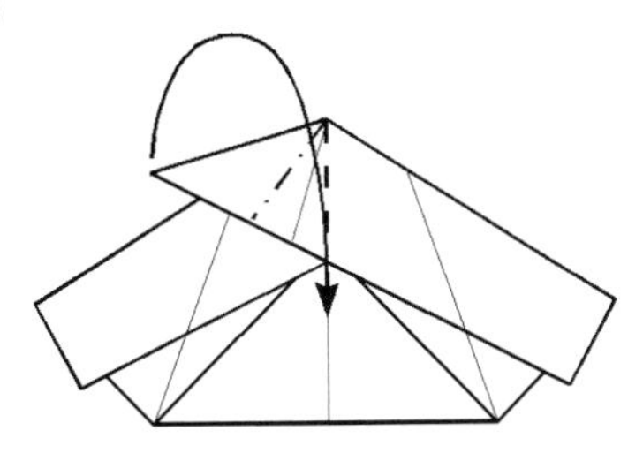

Lift the point upright, separate the layers and squash it flat so that the point ends up on the centre line.

19

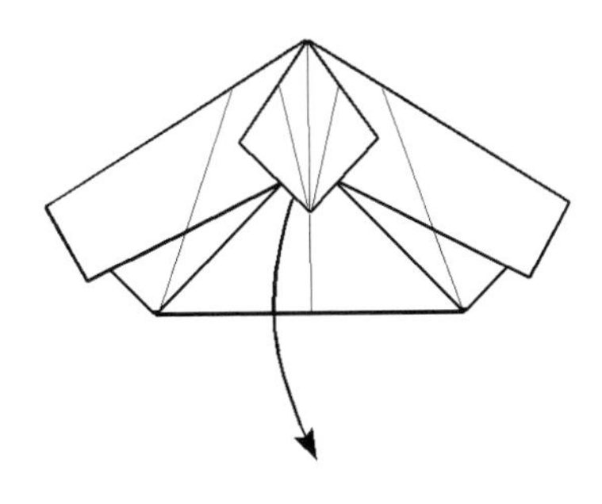

20

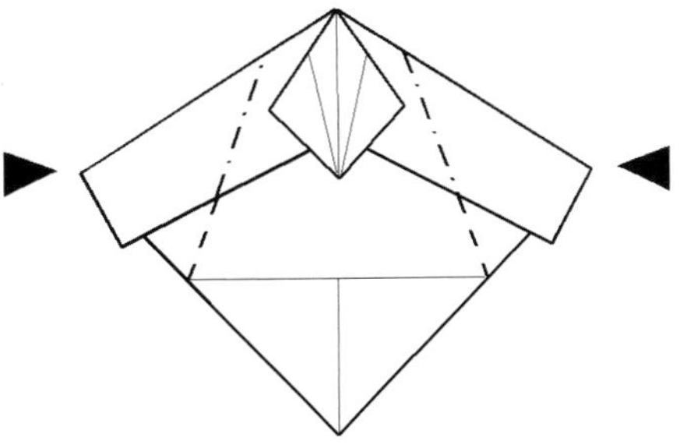

Fold the side flaps inwards between the front and back layers.

21

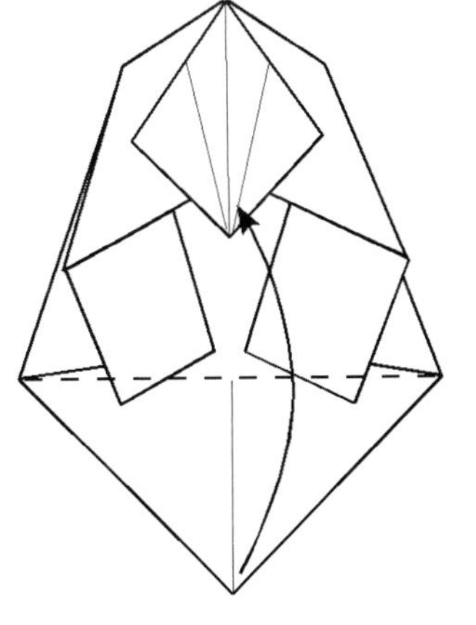

22

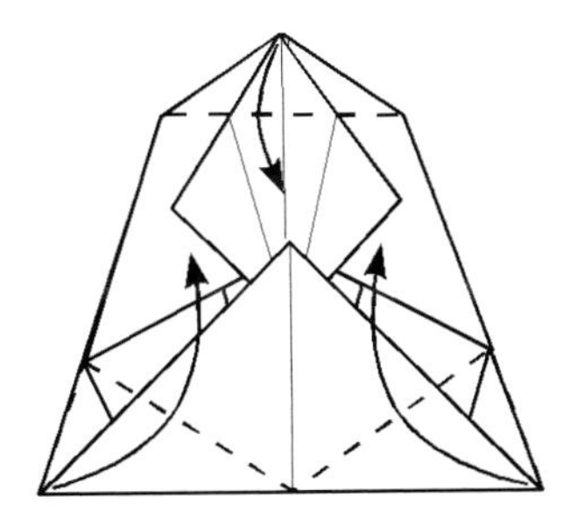

23

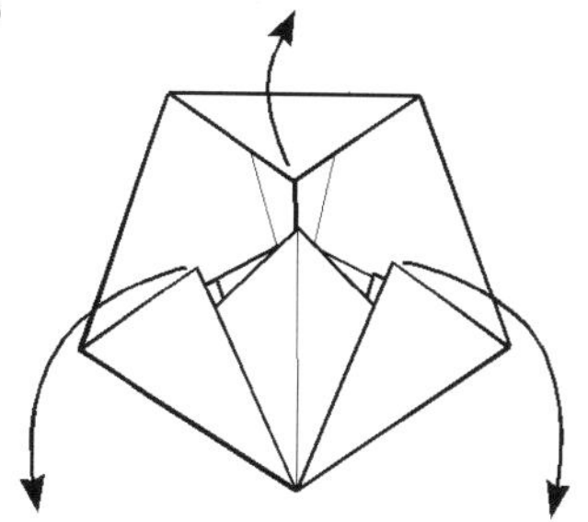

24

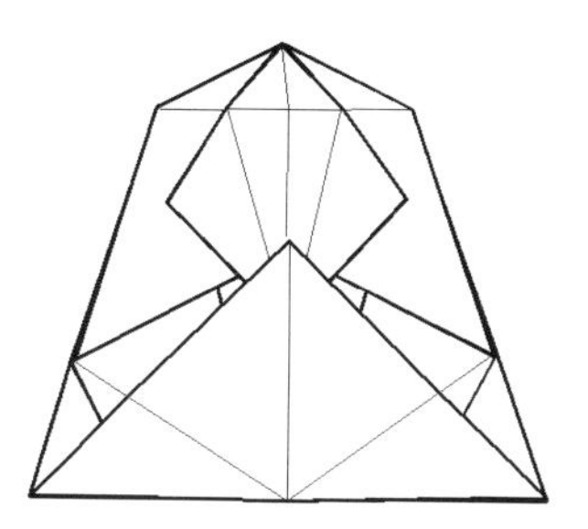

The modules fit together in male/female pairs.

This is the male version of the module.

To make the female form it is necessary to indent the upper flap.

First open the top of the module out and squash it flat as shown.

25

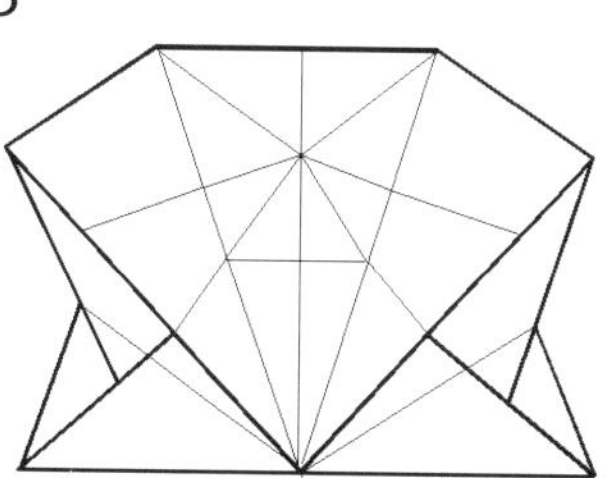

26

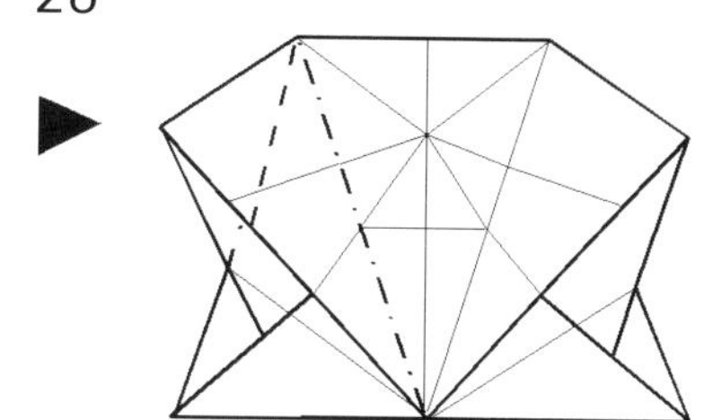

Lift up the left hand flap and tuck the edge under inside.

27

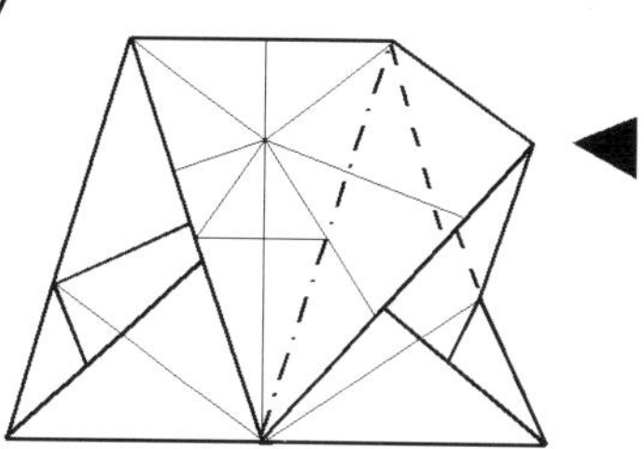

Repeat this manoeuvre on the right hand flap.

28

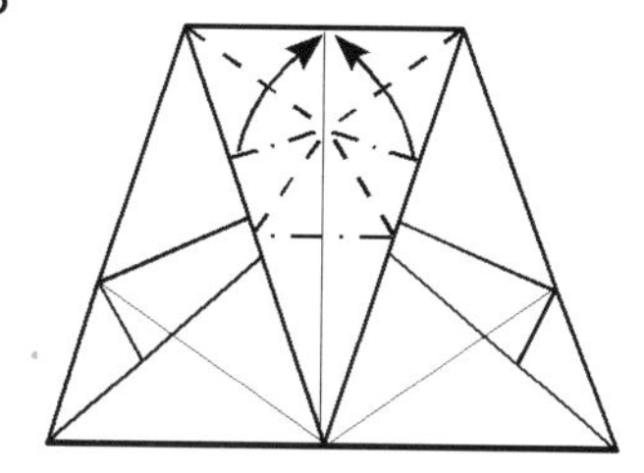

As you make these two small folds the tip of the point will widen. Squash it flat.

29

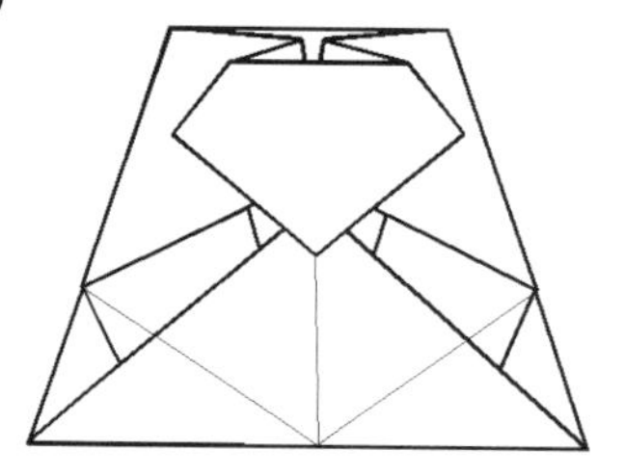

This is the female module.

30

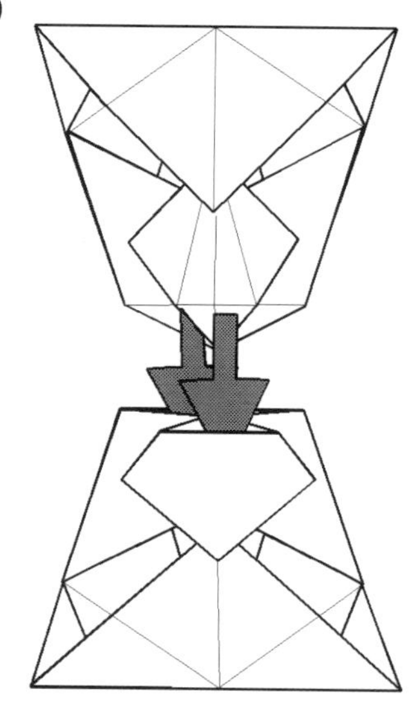

Note that there are two flaps and two sockets. Both flaps should be inserted simultaneously.

Turn over.

31

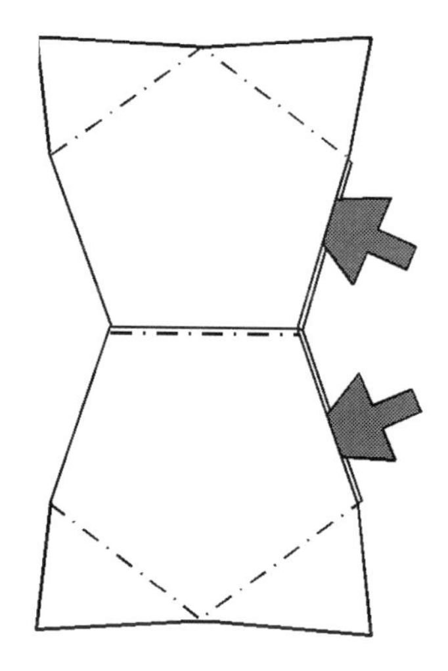

Two male/female pairs go together like this.

Each pair of modules will need to bend slightly at the join.

If the modules are folded accurately the flaps and pockets should be almost exactly the same width and depth.

Outline Dodecahedron

Eight sheets of A4 paper are required.

Begin by dividing each sheet into quarters. All thirty-two sheets should be folded to step 14, although only thirty modules are required.

1

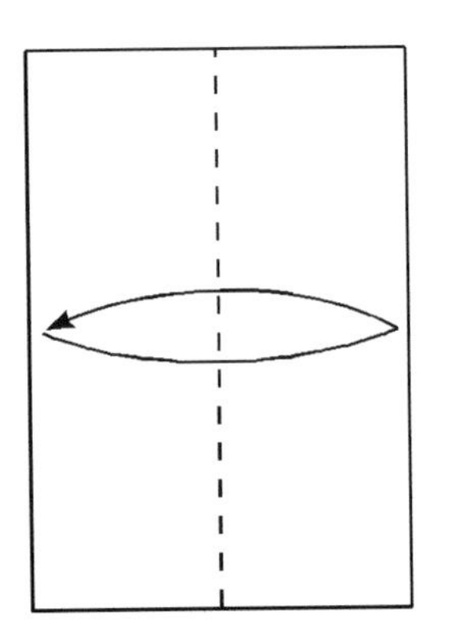

Turn over.

2

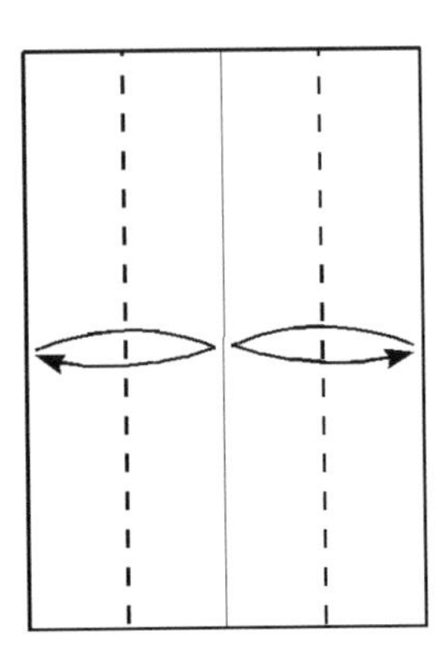

The module is easier to fold if each edge doesn't quite touch the centre line.

3

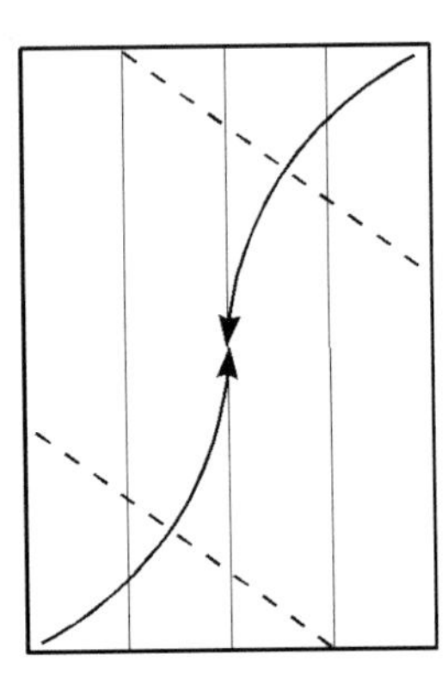

4

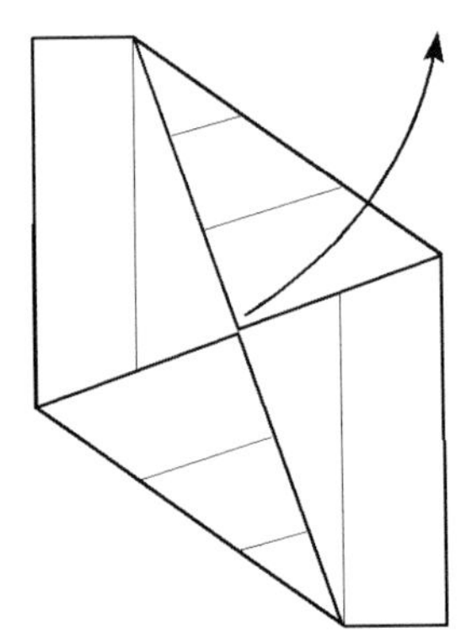

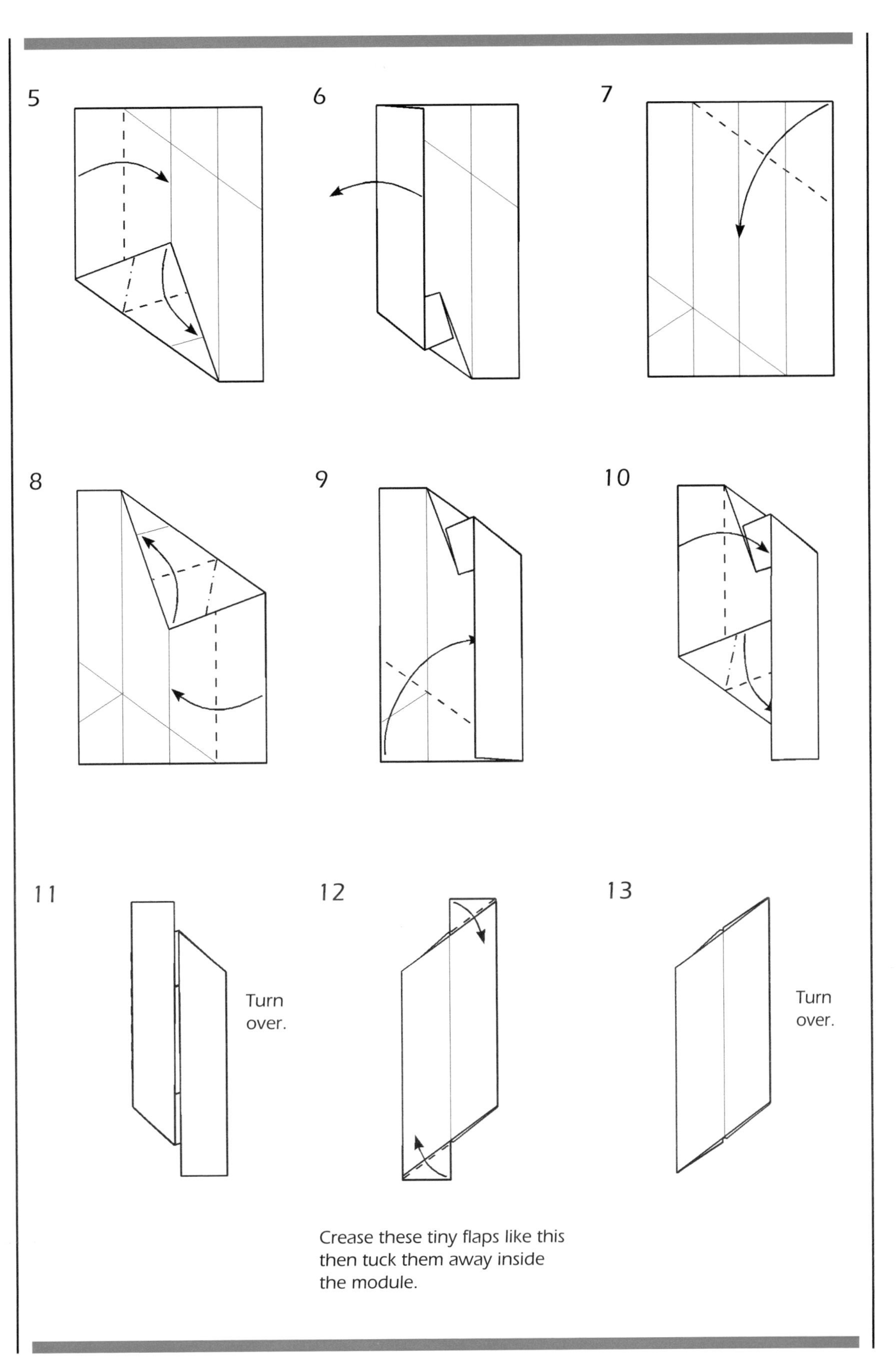
5
6
7
8
9
10
11
Turn over.
12
13
Turn over.
Crease these tiny flaps like this then tuck them away inside the module.

14

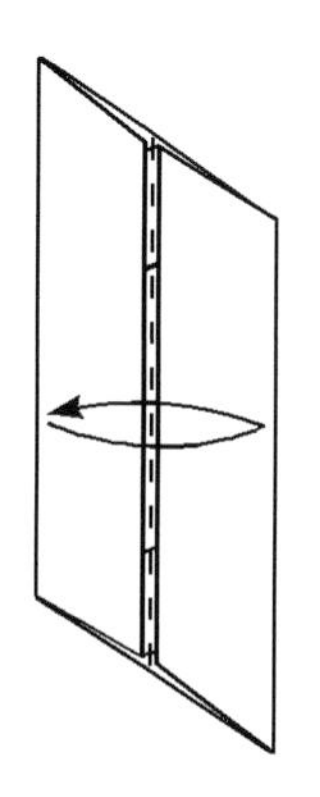

Make this crease lightly through all the layers.

Fold all 32 sheets to this stage.

15

Select two modules to use as your templates.

You should be able to fold 15 modules with each one before it wears out.

The template is shown shaded.

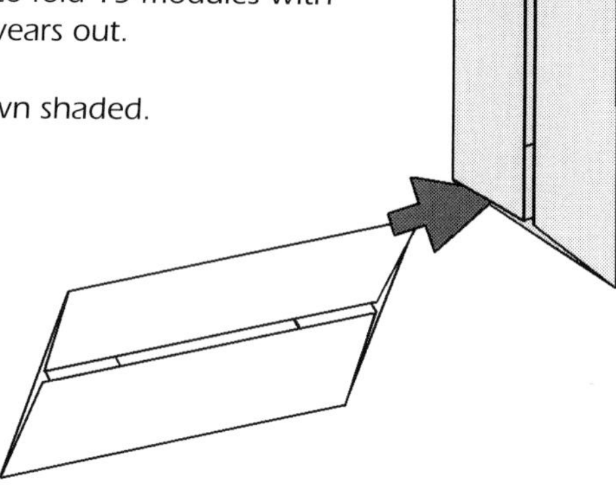

16

Crease firmly then take apart.

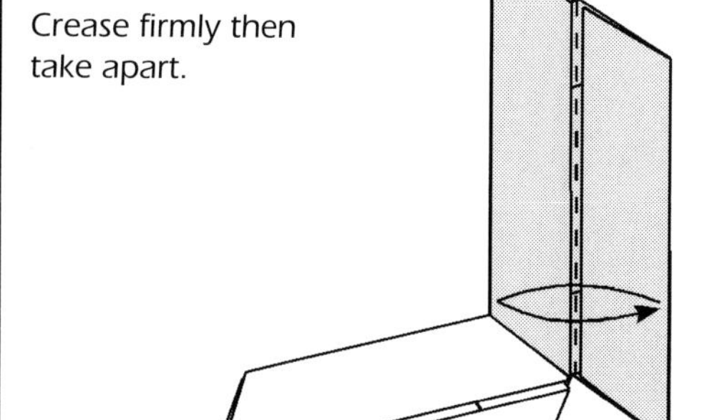

17 Repeat steps 15 and 16 on the other end of the same module.

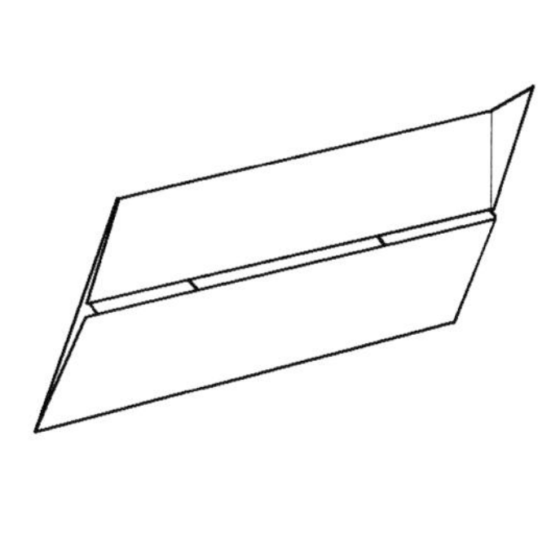

18

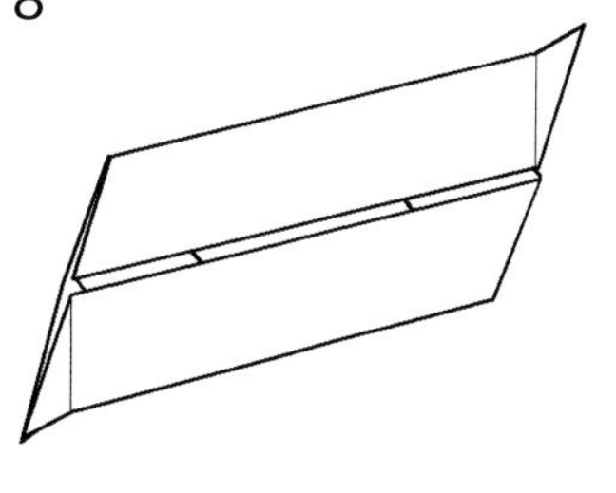

Turn over.

19

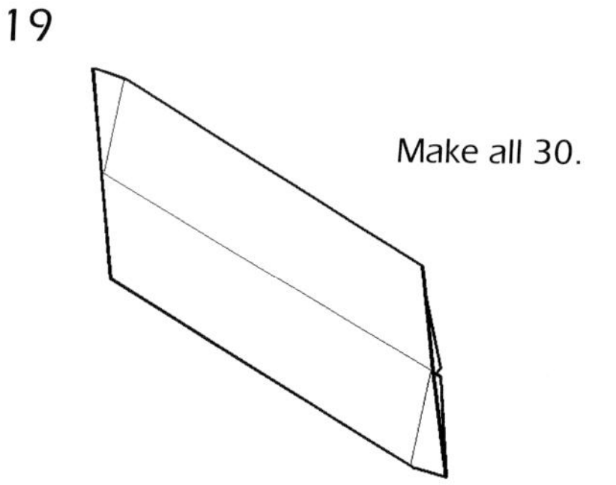

Make all 30.

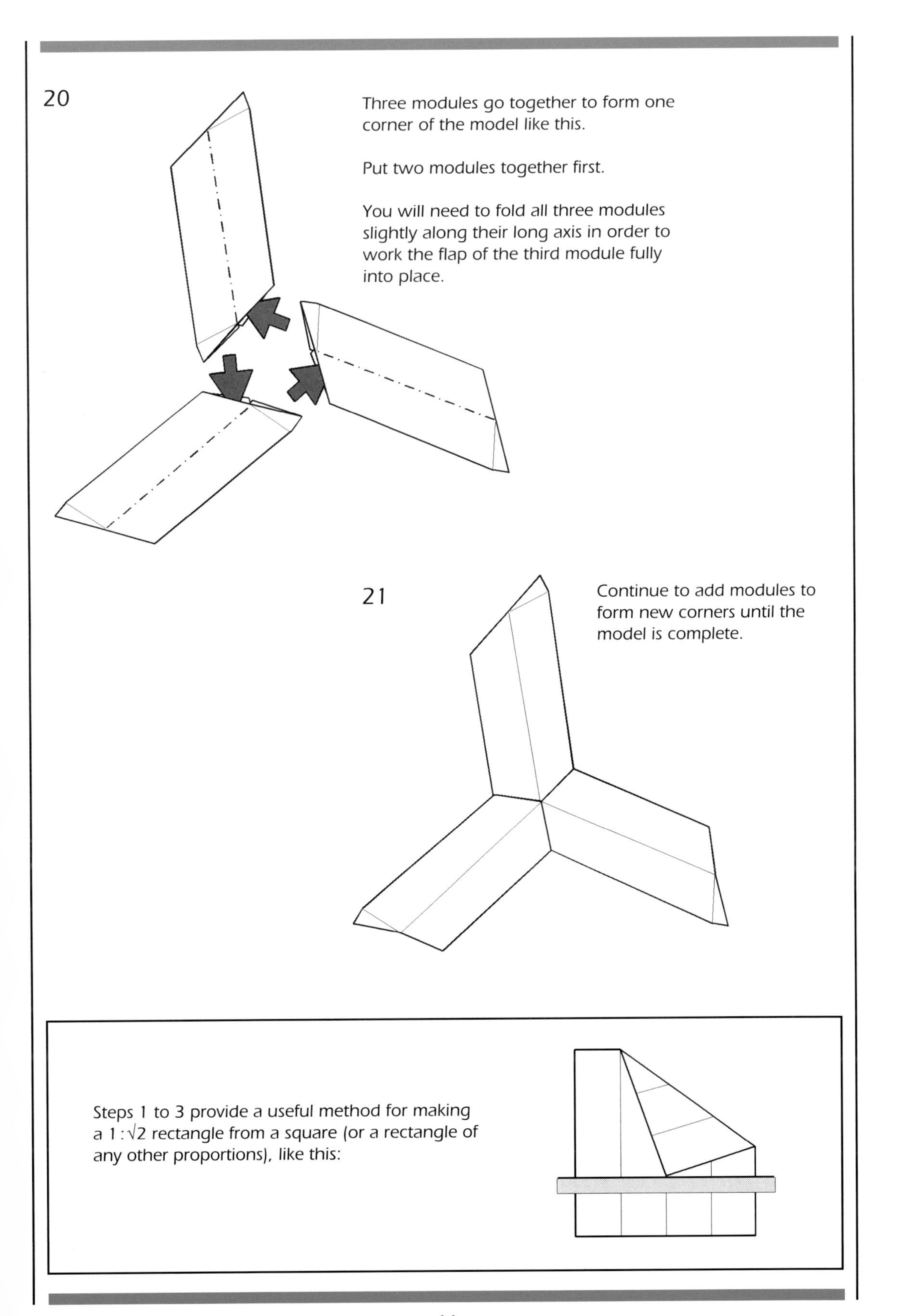

20

Three modules go together to form one corner of the model like this.

Put two modules together first.

You will need to fold all three modules slightly along their long axis in order to work the flap of the third module fully into place.

21

Continue to add modules to form new corners until the model is complete.

Steps 1 to 3 provide a useful method for making a 1 : √2 rectangle from a square (or a rectangle of any other proportions), like this:

Going further with modular origami

These models are only a small sample of the many wonderful designs that paperfolders have developed over the years. The best way to source other designs - and to keep up with new developments - is to join a paperfolding club. The one I belong to is called the British Origami Society - although it has members in over thirty countries across the world. Details of the society's activities and the benefits of membership can be obtained by contacting the British Origami Society via their website www.britishorigami.org.uk

Transformation

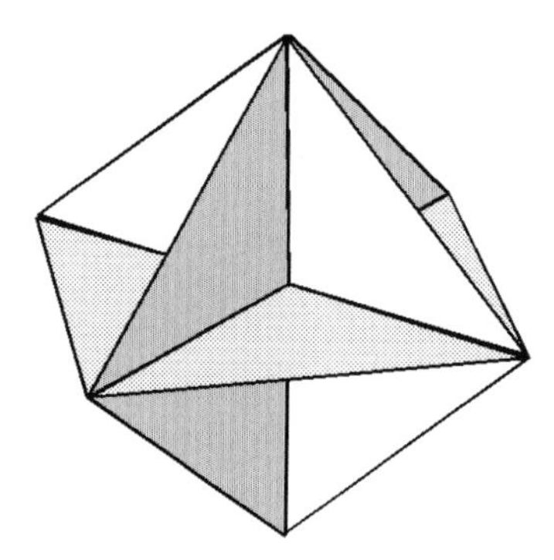

The Skeletal Octahedron described on page 28 is made up of six identical modules. Assembled in the way suggested there and illustrated above, you can see that the outer surface of the form can be regarded as a kind of distorted cube. All eight corners of a cube have been pushed inwards until they meet at the centre.

Once you appreciate this way of looking at it, take the six modules gently apart again. Strangely, it is possible to assemble these same modules into another, quite distinct, polyhedral form. This form is also a distortion of a cube, but the distortion is made in a different, less obvious way.

I leave you with this transformation as a puzzle to think about and to solve if you can. You do not have to add any more creases to any of the modules or to reverse the direction in which any of the existing creases lie.

What you will need, however, is a strong dose of lateral thinking!

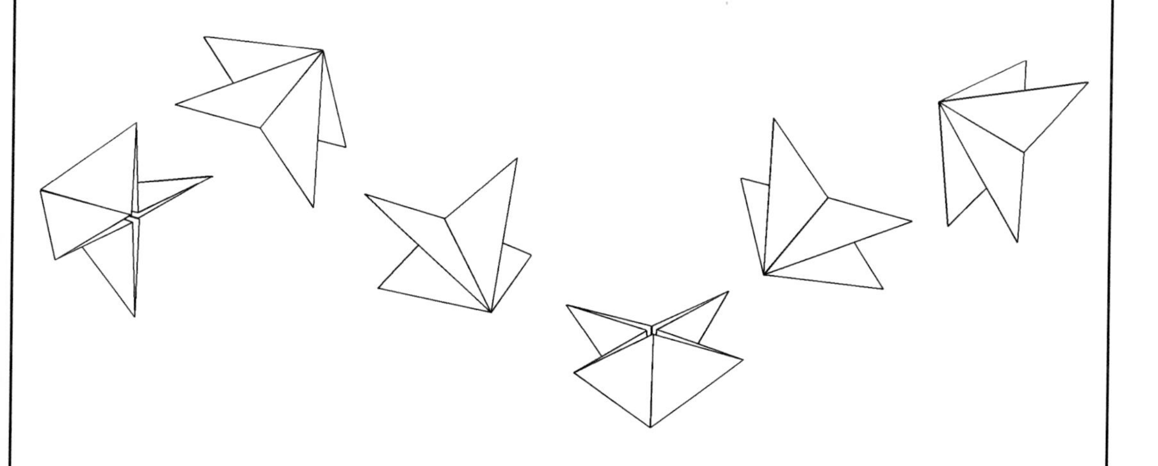

Bibliography

Anyone wishing to explore the mathematical aspects of origami further may also be interested in consulting the following books.

Origami For The Connoisseur by Kasahara and Takahama
Published in English by Japan Publications Inc. (I.S.B.N. 0 87040 670 1)
As well as some remarkable representational models, this book contains both single sheet and modular polyhedral folds by leading Japanese folders, many of whom are teachers or professors of mathematics.

Unit Origami by Tomoko Fuse
Published in Japanese by Chukuma (I.S.B.N. 4 480 04078 1)
A classic work that concentrates mainly on showing how folded modules can be used to divide the faces of each of the platonic polyhedra into patterned segments in various ways. Although the text is in Japanese, the diagrams are clearly drawn using standard international symbols and are easy to follow. This book may now be out of print and difficult to obtain.

Multi-Dimensional Transformations by Tomoko Fuse
Published in English by Japan Publications Inc. (I.S.B.N. 0 87040 852 6)
A varied collection of ingenious modular models diagrammed in Tomoko's usual clear and elegant style.

Brilliant Origami by David Brill
Published in English by Japan Publications Inc. (I.S.B.N. 0 87040 896 8)
Contains many wonderful modular folds as well as the author's amazing representational work, including a working nut and bolt.

The Encyclopedia of Origami and Papercraft Techniques by Paul Jackson
Published by Headline Book Publishing (I.S.B.N. 0 7472 0416 0)
Contains much material of interest about both modular and single sheet origami as well as other, more naturally organic, papercrafts.